Success guides

Standard Grade
Biology

Hannah Kingston ✕ Denyse Kozub

with contributions from George Milne

Contents

The Biosphere

The World of Plants

Animal Survival

Investigating Cells

The Body in Action

Inheritance

Biotechnology

Test Your Progress

Investigating an ecosystem

Biosphere/ecosystem/habitat

- The **biosphere** is the part of the planet where organisms are found.
- The biosphere can be divided into smaller units called **ecosystems**.
- The place where an organism (i.e. an animal or plant) lives is called a **habitat**.

If we want to learn about a particular habit, we need to carry out an investigation which involves:

1. collecting the organisms   **2. identifying** the organisms   **3. measuring** the physical conditions.

Sampling techniques

We cannot collect all the organisms within a habitat, so we must take a **sample**.

Using a quadrat

A **quadrat** is used to take **plant samples**.

A quadrat is a frame divided into squares. It is thrown **randomly** on the area being sampled. The number of squares which contain the plant being studied are counted, **not** the number of plants in each square.

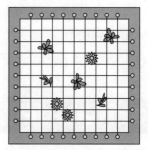

Using a pitfall trap

A **pitfall trap** is used to take **animal samples**.

The cup is placed in a hole, making sure that the top is level with the ground so that the animals fall in. However, some animals that fall in may be eaten by predators, such as spiders, which also fall in.

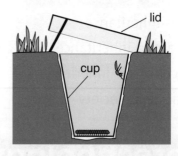

Identifying organisms

To identify organisms in a sample a **key** must be used.

A branching key

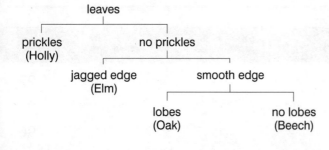

leaves

prickles (Holly)   no prickles

jagged edge (Elm)   smooth edge

lobes (Oak)   no lobes (Beech)

A paired statement key for leaves

1. with prickles . . . . .Holly
   with no prickles 2

2. jagged edge . . . . .Elm
   smooth edge 3

3. lobes . . . . .Oak
   no lobes . . . . .Beech

Measuring abiotic factors

Credit students need to identify sources of errors in different measuring techniques and know how to avoid them.

An **abiotic** factor is a **physical** factor, such as **light**, **moisture**, **oxygen concentration** or **temperature**.

Abiotic factors affect the organisms living in a particular habitat.

An abiotic factor can normally be measured using a meter.

Measuring moisture

A moisture meter is used. To avoid errors, you must be careful that you place the moisture probe firmly in the ground and wipe it dry afterwards.

Measuring light

A light meter is used. To avoid errors, you must be careful not to shade the meter and always hold it the same way when making the reading.

Effects of abiotic factors

Effect of abiotic factors	Reason why
Green plants are not found in areas with low light intensity.	Plants need light for photosynthesis.
Most land organisms are not found in very dry or very wet areas.	In very dry conditions they become dehydrated. In very wet conditions plant roots are deprived of oxygen.

Quick Test

1. What name is given to the place where an organism lives?

2. Name a sampling technique used to measure the level of abundance of plants.

3. Name a sampling technique used to measure the level of small invertebrate land animals.

4. Name the two types of keys used to identify organisms.

5. Name two abiotic factors.

6. What would you use to measure the moisture level in an environment?

7. State two precautions you would take to reduce errors when using a light meter.

Answers 1. Habitat. **2.** Using a quadrat. **3.** Using a pitfall trap. **4.** Branching, paired statement. **5.** Any two of light, moisture, temperature, oxygen concentration. **6.** Moisture meter. **7.** Be careful not to shade the meter; always hold it in the same way when making the reading.

How an ecosystem works

What is an ecosystem?

An **ecosystem** is made up of **living and non-living** things. All the parts of an ecosystem are **inter-related**.

Ecosystems can be divided into types, such as **coastal**, **mountain**, **river** or **forest** ecosystems.

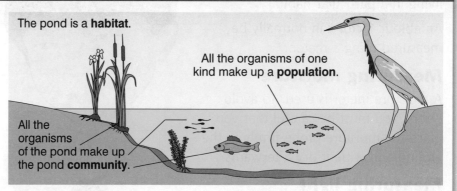

The pond is a **habitat**.

All the organisms of one kind make up a **population**.

All the organisms of the pond make up the pond **community**.

Community + habitat = ECOSYSTEM

Food and energy in an ecosystem

All living things need **energy**. They obtain their energy from food.
All the energy in an ecosystem comes from the sun, because **plants use light energy from the sun to make food.**

Producers and consumers

Producers – green plants use the sun's energy to produce food energy

Consumers – animals that get their energy from eating other living things

Primary consumers – animals that eat the producers

Secondary consumers – animals that eat the primary consumers

Tertiary consumers – animals that eat the secondary consumers

Herbivores – animals that only eat plants

Carnivores – animals that only eat animals

Top carnivores – animals that are not eaten by anything else except decomposers after they die

Food chains

- The way energy, in the form of food, passes from plants to animals and then to other animals can be shown by a **food chain**.
- The **arrows** in a food chain show the **direction of energy flow**.
- Food chains and webs **begin with energy from the sun**.

producer

primary consumer
herbivore
prey

secondary consumer
carnivore
predator

Food webs

- A food web gives us a more complete picture of who eats what.
- Most animals in a community eat more than one thing. If one kind of food runs out, they will be able to survive by eating something else.
- **Food webs are made up of many food chains linked together**.
- Food chains can be drawn for any environment.

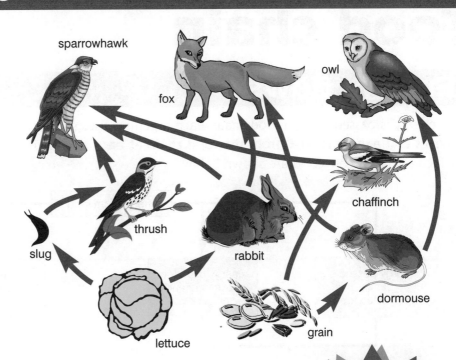

sparrowhawk · fox · owl · thrush · chaffinch · slug · rabbit · dormouse · lettuce · grain

What if?

- What would happen to a food web if an animal was removed by disease or other factors?
- In the food web, what would happen if the rabbit was removed?
  1. The amount of lettuce would increase at first as only the slugs will be eating it.
  2. The slugs would increase as there are more lettuces; the thrushes would increase at first because there are more slugs, but then the thrushes would get eaten by the sparrowhawks.
  3. The foxes and sparrowhawks would struggle at first with no rabbits, but then they would eat more dormice, chaffinches and thrushes.
  4. The number of owls, chaffinches and dormice would decrease.
- It all looks incredibly complicated, but you just have to think it through for every animal.
- **Look at who would get eaten, who would go hungry, what would they do about it and what effect it would have on the other animals in the food web.**

Top Tip
Learn all the definitions involved in food chains and webs.

Top Tip
In the exam you may be asked to extract a food chain from a food web.

Quick Test

1. Name the parts that make up an ecosystem.
2. Where does the energy in an ecosystem come from?
3. What is a producer?
4. What is a consumer?
5. Name the primary consumers in the food web.
6. What are carnivores?
7. What are herbivores?
8. How many top carnivores are there in the food web?
9. What do the arrows in food webs and food chains show?

Answers 1. Habitats and community. **2.** The sun. **3.** A plant that produces food from the sun's energy. **4.** An animal that eats other plants and animals. **5.** Slug, rabbit, chaffinch and dormouse. **6.** Animals that eat only other animals. **7.** Animals that eat only plants. **8.** Three **9.** The transfer of food energy between organisms.

Energy loss in a food chain

Energy flows

As energy is passed along a food chain, each organism uses some of it. This means that **energy is 'lost' from the food chain at each stage**. The ways in which energy is converted are shown below.

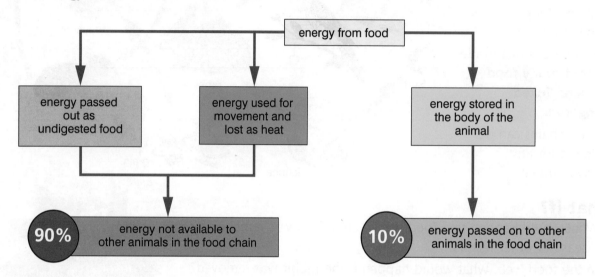

```
                        energy from food
           ┌───────────────┬───────────────────────────┐
           ▼               ▼                            ▼
   energy passed      energy used for          energy stored in
    out as            movement and             the body of the
  undigested food     lost as heat             animal
```

90% energy not available to other animals in the food chain

10% energy passed on to other animals in the food chain

Energy flow through a producer

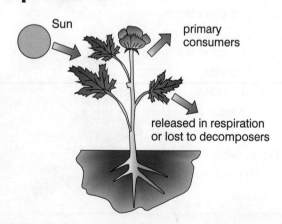

Sun

primary consumers

released in respiration or lost to decomposers

Energy flow through a consumer

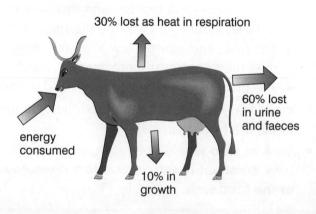

30% lost as heat in respiration

60% lost in urine and faeces

energy consumed

10% in growth

Pyramid of numbers

As you move along a food chain, very often the size of the organism increases but the number of them decreases.

| **grass** (very large number of small organisms) | → | **rabbit** (fewer, larger organisms) | → | **fox** (few, even larger organisms) | → |

This can be shown by drawing a pyramid.

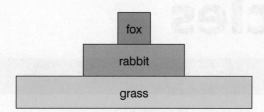

Sometimes a pyramid of numbers does not look like a pyramid at all as it does not take into account the **size** of the organisms.

In this pyramid of numbers the top carnivores are fleas that feed on single fox.

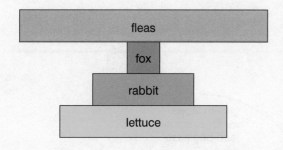

CREDIT

Pyramid of biomass

A more accurate idea of the **quantity of animal and plant material** in a food chain is obtained by constructing a **pyramid of biomass**. This represents the **mass of all the organisms at each level** and gives a much better representation of the actual quantity of animal and plant material at each level.

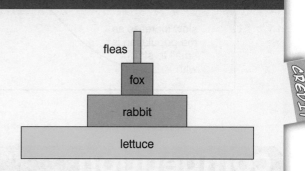

CREDIT

- Even though there are a lot of fleas, they weigh less than the fox they feed on.
- The fox weighs less than the number of rabbits it eats, and the number of lettuces the rabbits eat weigh more than the rabbits.

Top Tip
Make sure you know how energy flows in a food chain and reasons why energy is lost from a food chain.

Quick Test

1. Why do food chains only have four or five links?

2. What do pyramids of numbers show?

3. What pyramids can be drawn showing the mass of animals and plants?

4. Why do warm-blooded animals need to eat a lot of food?

5. In animals, where does most of the energy go?

6. List the ways that energy is lost in food chains.

Answers 1. Energy is lost at each link in the chain. **2.** The numbers of organisms at each level of a food chain. **3.** Pyramid of biomass. **4.** They need energy to keep warm. **5.** Lost in urine and faeces. **6.** Respiration, heat, waste and parts of the body not eaten.

Populations and nutrient cycles

Population growth

The sizes of most populations tend to stay roughly the same. The size of a population stays the same as long as the birth rate is the same as the death rate.

In most populations there is a **limit on numbers** which prevents a population explosion. Populations can be limited by:

a. predators

b. disease

c. limited food supply

d. lack of space, which may prevent breeding

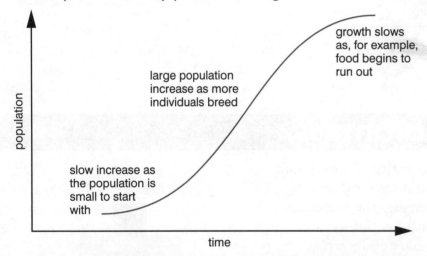

Competition

If different organisms eat the same food then **competition** will occur. Plants compete for light and water. Animals compete for food and a place to live. When competition occurs, some organisms will be more successful than others. These organisms will be more likely to survive.

Nutrient cycles

Bacteria and fungi are very important in an ecosystem. They feed on dead animals and plants and are known as **decomposers**.

Decomposers are important because:

a. they get rid of dead animals and plants

b. they **release chemicals** from dead organisms which go into the soil and help keep it **fertile**. These chemicals are taken up as nutrients by living organisms as part of nutrient cycles.

The nitrogen cycle

All living things need **nitrogen** to make **protein**.

Plants obtain nitrogen from the soil by taking in **nitrates**.

Animals obtain nitrogen by **eating plants** or **other animals**.

CREDIT

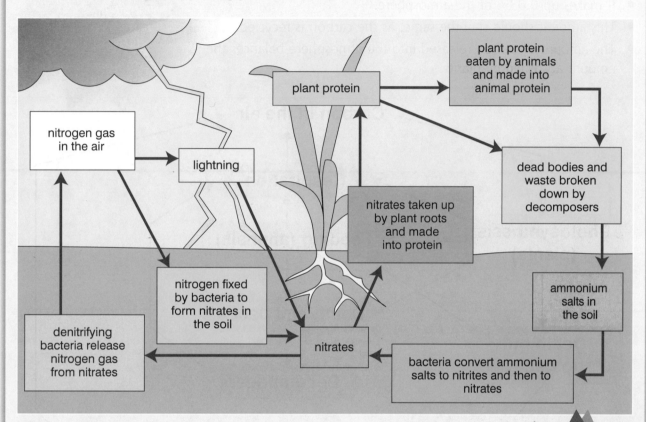

Nitrogen fixing means **absorbing nitrogen gas** from the atmosphere to make **nitrates**. Some bacteria in the soil can do this. Also, some plants (e.g. peas and clover) have swellings (nodules) on their roots in which these types of bacteria live.

Top Tip
Learn all the stages in the nitrogen cycle – questions on this topic are often badly answered.

Quick Test

1. Name two factors that influence the size of a population.

2. Name two factors that prevent a population explosion.

3. Name two things plants compete for.

4. What term is used to describe fungi and bacteria that feed on dead plants and animals?

5. Give two reasons why decomposers are important.

6. Why do living things need nitrogen?

7. In what form do plants obtain nitrogen from the soil?

8. Explain the term nitrogen fixing.

9. Name two types of plants that contain nitrogen fixing bacteria.

10. Name another way in which atmospheric nitrogen is changed into nitrates.

Answers 1. Birth rate, death rate. **2.** Predators, disease, limited food supply, lack of space. **3.** Light, water, root space. **4.** Decomposers. **5.** They get rid of dead animals and plants; they release chemicals from dead organisms which fertilise the soil; they are a vital link in nutrient cycles. **6.** To make proteins. **7.** Nitrates. **8.** It means absorbing the nitrogen gas from the atmosphere to make nitrates. **9.** Pea and clover plants. **10.** During lightning storms in the atmosphere.

The carbon cycle

The carbon cycle

- Carbon dioxide is a rare atmospheric gas.
- It makes up 0.03% of the atmosphere.
- This amount should stay the same, as the carbon is recycled.
- The amount of carbon released into the atmosphere balances the amount absorbed by plants.

Top Tip
The carbon cycle in the exam may look slightly different, so make sure you learn the processes involved.

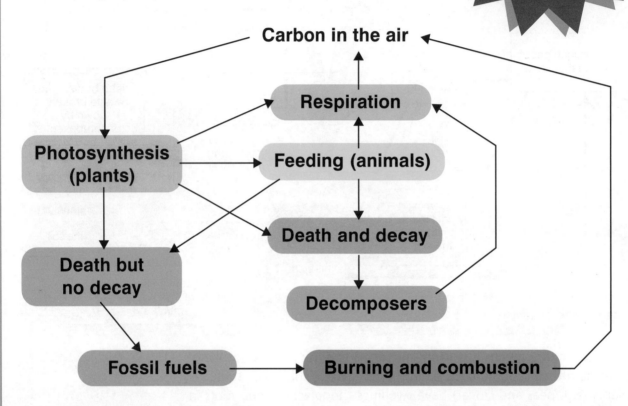

Carbon in the air

Respiration

Photosynthesis (plants)

Feeding (animals)

Death and decay

Death but no decay

Decomposers

Fossil fuels

Burning and combustion

Photosynthesis

Plants absorb carbon dioxide from the air. They use the carbon to make carbohydrates, proteins and fats using the **sun** as an energy source.

Feeding

Animals eat plants and so the carbon gets into their bodies.

Death and decay

Plants and animals die and produce waste. The carbon is released into the soil.

Decomposers

Bacteria and fungi present in the soil break down dead matter, urine and faeces, which all contain carbon.

Bacteria and fungi release carbon dioxide when they respire.

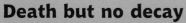

Death but no decay

Sometimes plants and animals die, but do not decay.

Heat and pressure gradually, over millions of years, produce fossil fuels.

Fossil fuels

Coal is formed from plants; oil and gas are formed from animals.

Burning and combustion

The burning of fossil fuels (coal, oil and gas) releases carbon dioxide into the atmosphere.

Respiration

Plants, animals and decomposers respire.

Respiration releases carbon dioxide back into the atmosphere.

Decomposition

- **Bacteria and fungi are decomposers**. They break down dead material.
- As well as helping to recycle carbon into the atmosphere, decomposers also recycle other **nutrients** into the soil (e.g. the nitrogen cycle on page 11).
- During photosynthesis plants take up these nutrients, which are dissolved in water.
- Animals eat plants; animals and plants eventually die, and the whole process begins again.
- Decomposition happens everywhere in nature, and also in compost heaps and sewage works.
- The ideal conditions for decomposition are **warmth, moisture and oxygen**.

Quick Test

1. Name the process that absorbs carbon dioxide from the air.

2. What are the two ways that carbon is released back into the air?

3. How do animals get the carbon into their bodies?

4. Name the three types of organisms that carry out respiration.

5. What is decomposition?

6. What organisms are involved in decomposition?

7. What happens to organisms that are not eaten and do not decay?

8. What are the ideal conditions for decomposition of dead matter to occur?

9. What do the plants do with the carbon they absorb?

10. What percentage of the atmosphere is carbon dioxide?

Answers 1. Photosynthesis **2.** Respiration and burning/combustion **3.** By eating plants or other animals **4.** Animals, plants and decomposers **5.** Breaking down of dead material **6.** Bacteria and fungi **7.** They are turned into fossil fuels **8.** Warmth, moisture and oxygen **9.** Turn it into carbohydrates, proteins and fats **10.** 0.03%

Control and management 1

Sources of pollution

Pollution is caused by the presence of a **substance that is harmful** to an animal or plant (e.g. oil in the sea.)

Pollutants come from three main sources: **industry**, **agriculture** and **domestic**.

Below are just some examples of pollutants that can affect fresh water, sea, land and air.

Industrial pollution

Energy sources and pollution

The energy industry produces electricity.

Most of our electricity comes from power stations which burn **fossil fuels** (coal, oil and gas).

Fossil fuels release harmful gases (such as sulphur dioxide and nitrogen dioxide) **into the air when they are burned**. When these gases dissolve in rainwater, they form **acid rain**.

- Acid rain kills fish and trees and damages buildings, particularly those made of limestone.
- Acid rain falls into lakes and poisons fish and birds.

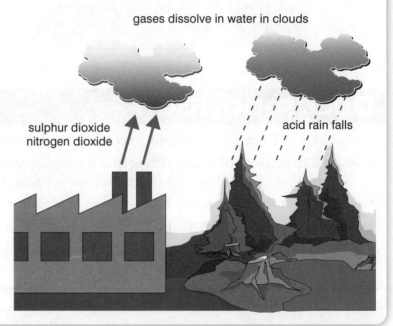

gases dissolve in water in clouds

sulphur dioxide
nitrogen dioxide

acid rain falls

Agricultural pollution

Crops are sprayed with pesticides.

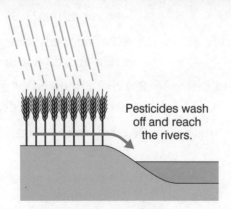

Pesticides wash off and reach the rivers.

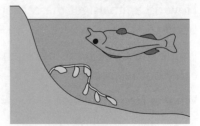

Pesticides are poisonous to many plants and animals, so the river habitat is damaged and many organisms are killed.

Pesticides used on farms can be washed into rivers and damage habitats and organisms.

Domestic pollution – sewage

Sewage is a common pollutant. Waste such as this is called **organic waste**. It can cause many changes in water environments if it is not treated at a sewage plant before being released into the wider environment.

Organic waste provides **food for bacteria** and allows them to grow and reproduce.

When bacteria feed on sewage, they **use up the oxygen in the water**. This means that there is less oxygen for other organisms such as fish and insects.

The chart below shows the changes that take place in a stretch of river polluted by sewage.

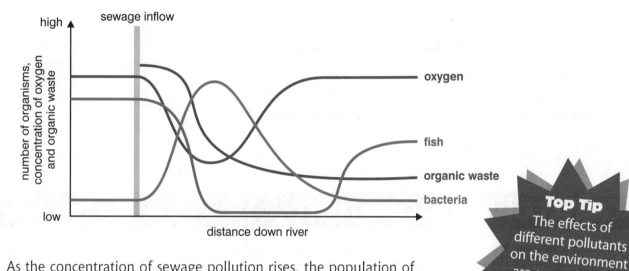

As the concentration of sewage pollution rises, the population of bacteria rises. This is because the bacteria feed off the sewage which provides raw materials and energy for growth and reproduction.

At the same time, the concentration of oxygen falls. This is because the bacteria use up the oxygen as they break down the organic waste in the sewage.

Consequently, animals, such as fish, stonefly nymphs and shrimps, decrease in numbers.

Top Tip
The effects of different pollutants on the environment are common exam questions.

Quick Test

1. Name the three main sources of pollution.
2. Name four parts of the environment that pollutants affect.
3. Name a gas which causes acid rain.
4. Name an agricultural chemical that can cause pollution.
5. How can we reduce the effect of domestic sewage on the environment?
6. What effect does domestic sewage have on the number of bacteria in a river?
7. How do the bacteria numbers affect the level of dissolved oxygen in the water?
8. How does the reduction in oxygen level affect the numbers of fish and invertebrate animals in the river?

Answers 1. Industry, agriculture, domestic. **2.** Air, fresh water, sea, land. **3.** Sulphur dioxide, nitrogen dioxide. **4.** Pesticides. **5.** Treat the sewage at a plant. **6.** The bacteria numbers increase. **7.** Oxygen levels decrease. **8.** Fish and invertebrate animal numbers decrease.

Control and management 2

Pollution indicators

- Some animals are only able to live in water that contains a lot of oxygen.
- Other animals can survive in water that contains little or no oxygen.
- The presence or absence of particular organisms can indicate whether the water is polluted or not.
- These animals are called **indicator species**.

Animals found in water with low levels of oxygen		
sludge worm	rat-tailed maggot	blood worm

Animals found in water with high levels of oxygen		
mayfly nymph	stonefly nymph	shrimp

Controlling pollution

It is essential that the activities of people are controlled so that pollution is reduced. Methods in place to control pollution include:

Pollutant	Method of control
Soot in smoke	Clean Air Act prohibits factories from releasing black smoke
Lead in exhaust fumes	introduction of unleaded petrol
Domestic sewage	treatment at sewage works before waste is discharged

Alternative sources of energy

The burning of fossil fuels in power stations to produce electricity pollutes the environment. Alternative sources of energy need to be used to reduce the harmful effects of fossil fuels.

Some alternatives to fossil fuels as sources of energy are:

a. **solar power** (using the sun's energy)

b. **wind power** (using the movement energy from wind)

c. **tidal power** (using the movement energy of tides)

d. **nuclear power** is also an alternative but it can be dangerous because it produces radioactive waste which can cause cancer and is also difficult to dispose of safely.

Management of resources

People have obtained many resources from the Earth (coal, oil, timber, food, etc.). This has resulted in the destruction or disruption of many habitats.

Conservation is very important because many resources will not last forever or may run short.

Three examples of poor management of resources and some possible solutions are:

Top Tip

Conservation and care of the environment is very topical and exam questions are often asked about this.

Poor management	Possible solution	
overfishing in the North Sea	have fish quotas, or increase net mesh size to allow smaller fish to survive	
destruction of rainforests	produce food more efficiently to limit the areas being logged for agriculture. Set aside areas as National parks, where forests cannot be damaged by land clearing for mining, logging or housing	NATIONAL PARK NO LOGGING
overuse of land, leading to desert soils	use different agricultural practices (crop rotation, natural fertilisers)	Organic Produce

Quick Test

1. What name is given to an organism whose presence or absence gives information about the level of pollution in a river?

2. Name two organisms that can live in water where the oxygen level is low.

3. Name two organisms that can only live in water which is unpolluted.

4. Why do we need to reduce the amount of fossil fuels burned for electricity?

5. Name two alternatives to fossil fuels as sources of energy.

6. Give two possible solutions to the problem of over fishing in the North Sea.

Answers 1. Indicator species. **2.** Sludge worms, rat-tailed maggot, bloodworm. **3.** Mayfly nymph, stone fly nymph, shrimp. **4.** Burning fossil fuels pollutes the environment. **5.** Solar power, wind power, tidal power. **6.** Increase net mesh size, have fishing quotas.

Introducing plants

The importance of plants

The sun is the source of all energy on Earth. **Plants are the link between the sun and other living things**.

Without green plants practically all life on Earth would not exist. People rely on a wide variety of plants, yet plant habitats are under constant threat.

Plants are required:

- for plant breeding **to produce new varieties**
- as **habitats** for other organisms
- as the initial **source of food** in a food web
- for **gas balance** in the atmosphere
- for **raw materials**, **food** and **medicines**
- for improving the **appearance** of our surroundings.

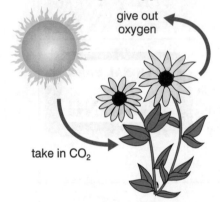

give out oxygen

take in CO_2

In the presence of light plants take in carbon dioxide and give out oxygen during photosynthesis. This helps to maintain the balance of gases in the atmosphere.

Green plants are the initial source of food in a food web. They convert the sun's energy into chemical (food) energy through photosynthesis.

The uses of plants

The range of use of plants is enormous. Here are some examples:

Foods	Raw materials	Medicines
wheat for bread	jute for string	foxglove for digitalis (a muscle relaxant)
palms for oil	flax for linen	
sugar cane for sugar	rose petals for perfume	poppy for morphine
grapes for wine	heather for dyes	cinchona tree for quinine

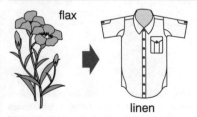

flax

linen

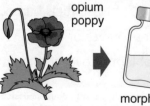

opium poppy

morphine

wheat

bread

CREDIT

The effects of reducing plant species

Plants, especially trees, are essential for maintaining life on this planet. However, some human activities are having very serious effects on the environment. The **destruction of habitats** (such as rainforests) means that many **species are being lost**. Other examples are:

- Selective breeding has resulted in a loss of certain plant species.
- Destruction of rainforests for farmland (e.g. The Amazon) has resulted in the extinction of many plant species.

New uses of plants

New techniques have allowed scientists to use plants to produce new products or to grow greater numbers of certain plants. For example:

- the extraction of protein (mycoprotein) from a fungus which can then be used as food
- the growing (culture) of oil palm cells to produce large numbers of individual plants from which palm oil can then be extracted.

Potential uses of plants

There are many species of plants that have not yet been discovered, which may have a potential use. But the destruction of habitats means that some of these plants will be lost forever.

Scientists are looking for plants that they can use to make new products. For example, the potential to use certain vegetable oils as fuels is being investigated. Some investigations can take a very long time and can be expensive in the short term. However, in the long term they could prove to be vitally important.

Top Tip
Make a list of the uses of plants.

Quick Test

1. Why are plants important in a food web?
2. How do plants help to maintain the balance of gases in the atmosphere?
3. Give three examples of foods that we get from plants.
4. Give three examples of raw materials that we get from plants.
5. Give three examples of medicines that we get from plants.
6. Give two effects of reducing plant species.
7. Name the food substance that scientists are now extracting from a fungus.
8. How might vegetable oil be used in the future?

Answers 1. They are the initial source of food in a food web, because they convert the sun's energy into chemical/food energy in photosynthesis. **2.** Plants take in carbon dioxide and give out oxygen. **3.** Bread from wheat (flour); sugar from sugar cane; wine from grape juice. **4.** Jute for string; flax for linen; rose petals for perfume; heather for dyes. **5.** Digitalis (a muscle relaxant) from foxglove; morphine from poppy; quinine from cinchona tree. **6.** Animal habitats may be destroyed; animal food sources may be lost. **7.** Mycoprotein. **8.** As a source of fuel.

Plant growth

The life cycle of a plant

In order to reproduce, some plants produce **seeds**. This involves a number of stages in the plant life cycle.

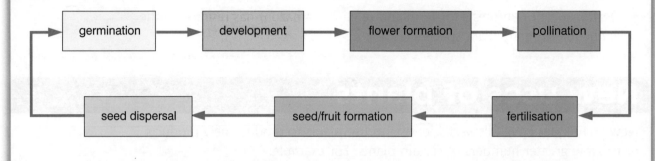

```
germination → development → flower formation → pollination
                                                      ↓
seed dispersal ← seed/fruit formation ← fertilisation
```

Seed structure and germination

Germination is the development of a new plant from the embryo plant in a seed. Seeds need **water**, **oxygen** and the **right temperature** before they will germinate.

- **Water** is needed to activate enzymes which digest stored food.
- **Oxygen** is needed for the production of energy for germination.
- **Warmth** is needed for the enzymes involved in germination to work effectively.

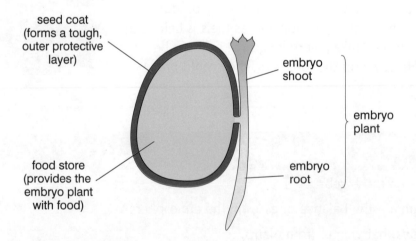

seed coat (forms a tough, outer protective layer)

embryo shoot

embryo plant

food store (provides the embryo plant with food)

embryo root

Changes in percentage germination at different temperatures

- At very high (above 45°C) and very low (below 5°C) temperatures, seed germination is often poor.
- Seeds normally have a high percentage germination between 15°C and 30°C.
- The best temperature for germination of a species of plant is known as the **optimum temperature**.

CREDIT

The structure of a flower

To **reproduce sexually** some plants produce **flowers**. The flowers contain the **sex organs**.

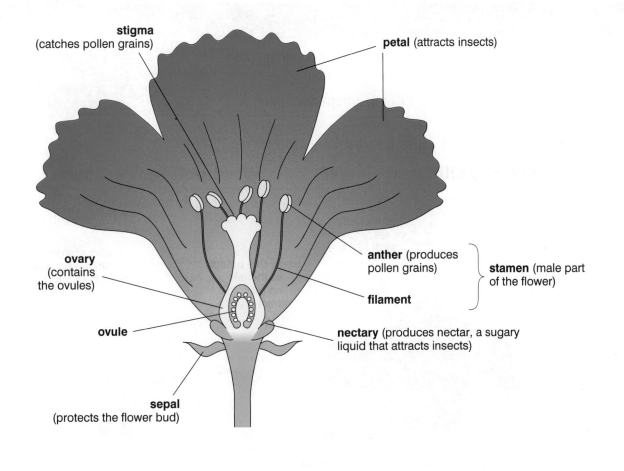

stigma (catches pollen grains)

petal (attracts insects)

ovary (contains the ovules)

ovule

anther (produces pollen grains)

filament

stamen (male part of the flower)

nectary (produces nectar, a sugary liquid that attracts insects)

sepal (protects the flower bud)

Quick Test

1. Why is the seed coat important?
2. Name the parts of the embryo plant.
3. Why is the food store in a seed important?
4. Name three factors required for germination.
5. Where are a plant's sex organs found?
6. Name two parts of a flower that attract insects?
7. Name the male part of a flower.
8. Which part of the flower contains the ovules?
9. Which part of the flower produces pollen?
10. Which part of the flower catches pollen grains?

Top Tip
Learn all the parts of the flower because you may be asked to fill in the missing labels on a diagram.

Pollination and fertilisation

Pollination

A new seed is formed when the male sex cell in a pollen grain joins up with a female sex cell. Sex cells are called **gametes**.

Pollination involves the transfer of pollen from the anther to the stigma.

Many plants are pollinated by insects or by the wind.

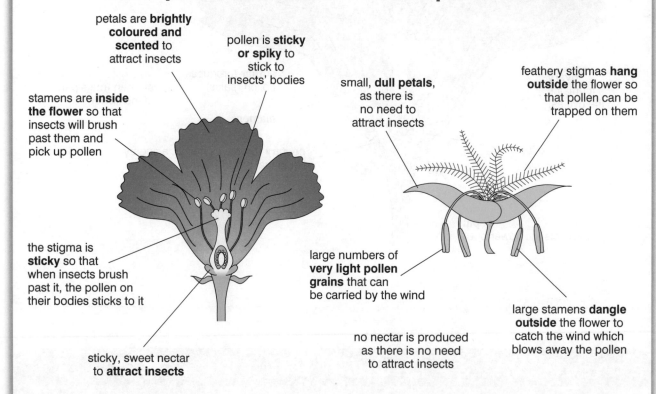

Insect pollinated flower

petals are **brightly coloured and scented** to attract insects

pollen is **sticky or spiky** to stick to insects' bodies

stamens are **inside the flower** so that insects will brush past them and pick up pollen

the stigma is **sticky** so that when insects brush past it, the pollen on their bodies sticks to it

sticky, sweet nectar to **attract insects**

Wind pollinated flower

small, **dull petals**, as there is no need to attract insects

feathery stigmas **hang outside** the flower so that pollen can be trapped on them

large numbers of **very light pollen grains** that can be carried by the wind

large stamens **dangle outside** the flower to catch the wind which blows away the pollen

no nectar is produced as there is no need to attract insects

Fertilisation

After the pollen grain has landed on the stigma, fertilisation occurs. **Fertilisation involves the fusion of the male gamete and the female gamete.**

Once fertilisation has taken place, the ovule becomes a seed and the ovary becomes a fruit. The petals then die and drop off.

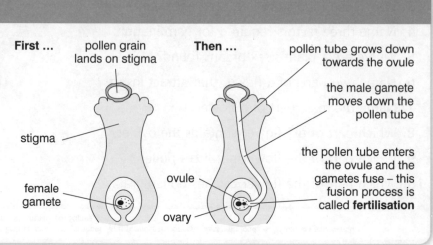

First ...
pollen grain lands on stigma

stigma

female gamete

Then ...
pollen tube grows down towards the ovule

the male gamete moves down the pollen tube

the pollen tube enters the ovule and the gametes fuse – this fusion process is called **fertilisation**

ovule

ovary

Seed dispersal

Plant seeds are contained in the fruit. There are different kinds of fruit, for example:

a. Fleshy fruits (e.g. tomato, plum, apple). The main part of fleshy fruit is soft and juicy.

b. Dry fruits (e.g. dandelion, sycamore). The main part of dry fruit is hard and dry.

Seeds must be carried away from the parent plant to reduce overcrowding and competition for water, light and nutrients.

Seeds can be dispersed in different ways, for example:

Dispersal method	Description of some examples	Seeds/fruits
wind	– seeds may have extensions which act as parachutes or wings to carry the seed in the wind (e.g. dandelion and sycamore) – fruits may also be shaken like a pepper pot (e.g. poppy)	
animal (external)	– carried away by animals and dropped (e.g. hazelnuts) – have hooks which attach to the animal's fur and may be rubbed off later (e.g. burdock)	
animal (internal)	– brightly coloured fruits attract animals. When eaten, the seed (e.g. cherry, tomato) isn't digested and is dropped by the animal or passed out in the faeces	

Quick Test

1. What term is used to describe male and female sex cells?

2. What is pollination?

3. Name the two types of pollination.

4. Describe the differences in the pollen between insect and wind pollinated flowers.

5. Explain why the stigmas of wind pollinated flowers hang outside the flowers.

6. Why do wind pollinated flowers not produce nectar?

7. How does the male gamete reach the female gamete?

8. What is a fruit?

9. Name three ways in which fruits and seeds are dispersed.

10. Why is it important that fruits and seeds be carried away from the parent plant.

Answers: 1. Gametes **2.** It is the transfer of pollen from the anther to the stigma. **3.** Insect pollination; wind pollination. **4.** Insect pollinated flowers – pollen is sticky or spiky; wind pollinated flowers – pollen is light. **5.** So that they are in a position where pollen can land on them. **6.** They do not need to attract insects. **7.** It passes down the pollen tube. **8.** Any part of a plant that contains seeds. **9.** By wind; animal (external); animal (internal). **10.** They would be too crowded; competition would take place for water, light and nutrients.

Asexual reproduction

One parent

Sexual reproduction involves two parents. Many flowering plants can reproduce in a way that involves only one parent. This is called **asexual reproduction** and does not involve the formation of sex cells.

There are many ways in which plants can reproduce asexually. Three examples are:

Runners

A runner is a side shoot which grows out from the parent. Where it touches the ground a new plant grows (e.g. strawberry, spider plant).

Tubers

Food made in the leaves is stored in tubers and used as an energy source for new growth (e.g. potato, dahlia).

Bulbs

A bulb has thick fleshy leaves full of stored food which is used for the growth of a new plant the following year (e.g. daffodils, onions).

Comparing asexual and sexual reproduction

	Asexual reproduction	Sexual reproduction
Advantages	– early quick growth is possible because there are no vulnerable stages involved – offspring will all have the parent plant's good characteristics	– variation takes place, which may be an advantage if conditions change – reduces competition arising from dense growth around the parent – allows dispersal of seeds to new areas
Disadvantages	– because there is no variation, weak characteristics can be passed on – overcrowding can take place	– involves many vulnerable stages which the young plant may not survive

Asexual reproduction involves the formation of **clones**. A clone is a group of cells or organisms all originating from the same parent and all **genetically identical** to each other and the parent.

Top Tip
Learn the advantages and disadvantages of asexual reproduction and compare them with sexual reproduction.

Artificial propagation

Gardeners can make use of a plant's ability to reproduce asexually by using different methods of **artificial propagation**. Instead of growing seeds, they use a small section of stem, root or leaf to grow a plant genetically identical to the parent plant.

Three common methods of artificial propagation are:

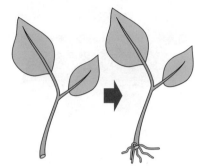 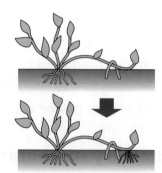

Cuttings

Cuttings are **small pieces of stem** cut from a healthy plant. The cut stem can be placed in a rooting medium to encourage root growth (e.g. geranium).

Grafting

A portion of a plant with good flower or fruit growth is taken and joined to a plant with an established, strong root system (e.g. roses and fruit bushes).

Layering

The **stem** of the parent plant is **bent** until it touches the ground. It is then held in place until roots have formed (e.g. carnations).

Commercial advantages of artificial propagation

Artificial propagation has brought enormous benefits to both agriculture and horticulture.

a. It is a quick method of **producing large numbers** of genetically identical new plants.

b. Particular varieties that are required can be produced easily.

c. Techniques, such as grafting, produce a plant that will grow fruit or flowers of a known variety or quality.

CREDIT

Quick Test

1. How many parents are involved in asexual reproduction?

2. Name three ways in which plants reproduce asexually?

3. Give two examples of plants that reproduce asexually by producing runners.

4. What term is used to describe a population of genetically identical plants produced by asexual reproduction?

5. Name three common methods of artificial propagation.

6. Give the commercial advantages of artificial propagation.

Answers: 1. One **2.** Runners, tubers, bulbs. **3.** Strawberry, spider plant. **4.** Clones. **5.** Cuttings, grafting, layering. **6.** It is a quick method of producing large numbers of new plants of known variety or quality.

Making food

How plants make food

Plants support all food chains and webs because they manufacture food at the start of food chains.

Plants use **light energy** to make food from **carbon dioxide** and **water**. This process is called **photosynthesis**.

The chemical **chlorophyll** (which gives plants their green colour) **traps the light energy** from the sun.

This **light energy is then converted to chemical energy** in the form of glucose.

Top Tip
Make sure you know the raw materials required for photosynthesis and the products of photosynthesis.

A summary of photosynthesis

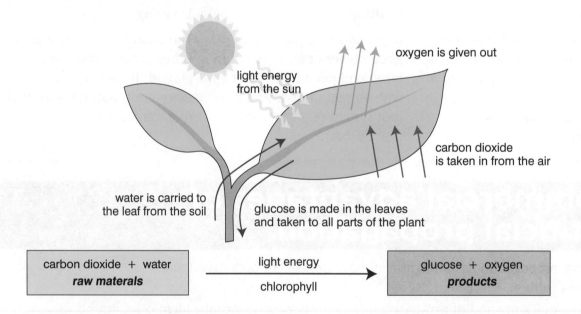

| carbon dioxide + water
raw materals | light energy
——————————————→
chlorophyll | glucose + oxygen
products |

The leaf takes in **carbon dioxide** through tiny pores on the surface of the leaf. A single pore is called a **stoma** (plural – **stomata**). These pores can open and close. Oxygen and water vapour are lost from the leaves through the stoma.

The use of glucose by plants

The glucose manufactured during photosynthesis is used by plants in a number of ways.

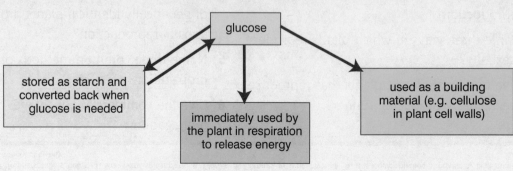

CREDIT

Transport in a plant

From soil to xylem

- **Water** and **dissolved minerals** enter the root cells by **osmosis**.
- The **root hairs** on the surface of the root provide a **large surface area** for absorption.
- Osmosis causes the water to move from cell to cell until it reaches the xylem vessels.

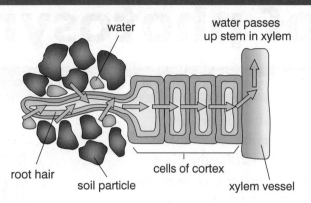

water

water passes up stem in xylem

root hair

soil particle

cells of cortex

xylem vessel

From xylem to leaves

Xylem carry water and **minerals** from the soil to the leaves for photosynthesis.

CREDIT

The xylem are dead cells joined together to make up tubes. They have thick strong walls made of **lignin**, which give the plant support.

The water moves up the stem to the leaves in the transpiration stream.

Phloem

Phloem cells **carry sugars** from the leaves up and down the plant to where it's needed.

All phloem cells are alive. The end walls have pores. Food is transported through the pores from cell to cell.

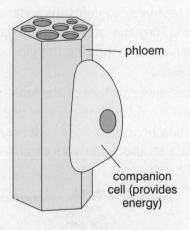

phloem

companion cell (provides energy)

CREDIT

Vascular bundle

Phloem and **xylem** vessels often run side by side and form a **vascular bundle**.

Vascular bundles are found in the **roots**, the **stem** and in the **veins** of a leaf.

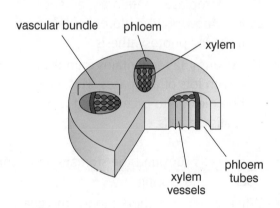

vascular bundle

phloem

xylem

xylem vessels

phloem tubes

Quick Test

1. What does a plant need for photosynthesis?

2. What does a leaf produce during photosynthesis?

3. What is chlorophyll?

4. How do the leaves obtain water?

5. How does the plant obtain carbon dioxide?

6. List three uses of the glucose produced by photosynthesis.

7. How is glucose carried from the leaves to every part of the plant?

8. Name the storage form of carbohydrate in a leaf.

Answers. 1. Carbon dioxide, water, chlorophyll, sunlight. **2.** Oxygen and glucose. **3.** A green pigment that absorbs energy from the sun. **4.** Through the roots and xylem tubes up to the leaf by osmosis. **5.** From the air through the stomata. **6.** Used to make cellulose, stored as starch, used up during respiration to release energy. **7.** Transported in the phloem tubes. **8.** Starch.

The leaf – the organ of photosynthesis

Top Tip
Make sure you understand how the design of a leaf is related to its function.

CREDIT

The leaf structure

The leaf is the **organ of photosynthesis**. It makes all the food for the plant.

- The upper surface of the leaf is called the **waxy cuticle**. It is a waterproof layer that cuts down the loss of water by evaporation.
- The upper cells of the leaf make up the **epidermis**. Light passes straight through these. The next layer of cells contains the **palisade cells**. This is where most photosynthesis takes place. These contain lots of **chloroplasts**. The chloroplasts contain a pigment called **chlorophyll**. Chlorophyll absorbs sunlight for photosynthesis.
- The **spongy layer** contains rounded cells with lots of air spaces. This allows carbon dioxide to circulate and reach the palisade cells.

waxy cuticle

epidermis

palisade cell (contains chloroplasts)

spongy mesophyll cells

leaf vein

guard cell   stoma

- The **leaf vein** contains the **xylem and phloem** tubes. They run over the plant, supplying it with water and transporting the glucose that is produced.
- At the lower surface of the leaf are tiny pores called **stomata** (one pore = one stoma). The stomata open and close to let carbon dioxide diffuse in and water vapour and oxygen out.

- Guard cells surround the stomata and control their opening and closing.
- When water is in short supply, the guard cells become **flaccid** and less curved; this closes the stomata and prevents water being lost from the leaf.
- When a plant has plenty of water the guard cells become **turgid** and curved; the stomata are open and water can escape from the leaf.

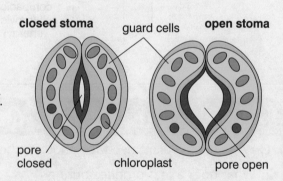

closed stoma   guard cells   open stoma

pore closed   chloroplast   pore open

Key Facts

The leaf has many features that enable it to carry out photosynthesis in the best possible way. These include:
- Leaves are **flat** with a **large surface area** to absorb as much sunlight as possible.
- They are **thin**, so carbon dioxide can reach the inner cells easily.
- They have **plenty of stomata** in the lower skin.
- They have **plenty of veins** to **support** the leaf and **carry substances** to and from all the cells in the leaf and plant.

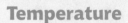

Factors affecting the rate of photosynthesis

We can measure the rate of photosynthesis by how much oxygen is produced in a given time. There are three things that affect the rate of photosynthesis. We call them **limiting factors**. They are:

Light

❶ If the light intensity is increased, photosynthesis will increase steadily, but only up to a certain point.

❷ After this point, increasing the amount of light will not make any difference, as it will be either the amount of carbon dioxide or the temperature that is the limiting factor.

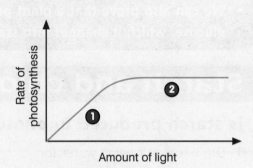

Temperature

- You can increase light and carbon dioxide to increase photosynthesis as much as possible, but the temperature must not get too hot or too cold.

- Usually the rate of photosynthesis is limited by the temperature being too low, as is the case for plants not normally grown in Britain.

- Greenhouses help maintain a high enough temperature for optimum growth conditions.

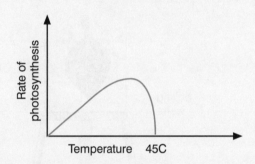

Carbon dioxide

❶ If the carbon dioxide concentration is increased, photosynthesis will increase up to a certain point.

❷ Beyond that point, light or temperature become the limiting factor.

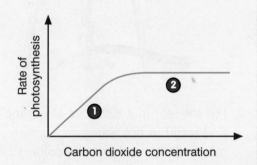

CREDIT

Quick Test

1. Give two structural features of a leaf that make it a good design for photosynthesis.

2. Why do you think that the palisade cells are near the surface of the leaf?

3. Name the cells that surround the stomatal openings.

4. The spongy mesophyll cells are loosely arranged. Explain the importance of this.

5. What three factors limit the rate of photosynthesis.

Answers. 1. Large surface area, flat, thin, stomata and veins **2.** So that they can absorb as much sunlight as possible through their chloroplasts. **3.** Guard cells. **4.** The large spaces between the cells allow gases to diffuse quickly. **5.** Amount of light, carbon dioxide and temperature.

Photosynthesis experiments

- We can prove that a plant needs chlorophyll, light and carbon dioxide for photosynthesis.
- We can also prove that a plant produces oxygen and glucose, which it changes into starch.

Top Tip
How to test a leaf for starch is an important method to learn.

Starch and chlorophyll

Is starch produced in photosynthesis?

1. Dip a leaf in boiling water for about a minute to soften it.

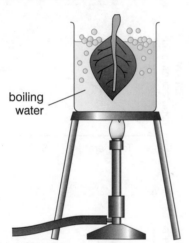

boiling water

2. Put the leaf in a test-tube of ethanol and stand in hot water for 10 minutes (this removes the colour).

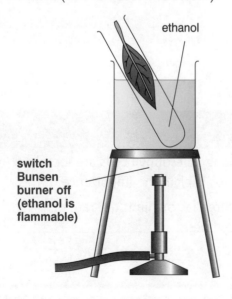

ethanol

switch Bunsen burner off (ethanol is flammable)

3. Remove and wash the leaf.

4. Lay the leaf flat in a petri dish and add **iodine** solution.

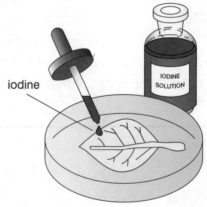

iodine

IODINE SOLUTION

5. If starch is present the leaf should go blue/black.

- This experiment can also prove that **chlorophyll** is needed if you use a variegated leaf. A variegated leaf has some white parts where there is no chlorophyll.

a variegated leaf

- The leaf should only go black where the leaf was green.

Light, CO$_2$ and oxygen

Does a plant need light?

- De-starch a plant by leaving it in the dark for 24 hours.

- Cover part of the leaf with some foil and leave in the sun for a few hours.

- Test the leaf for starch; only the uncovered part of the leaf will contain starch.

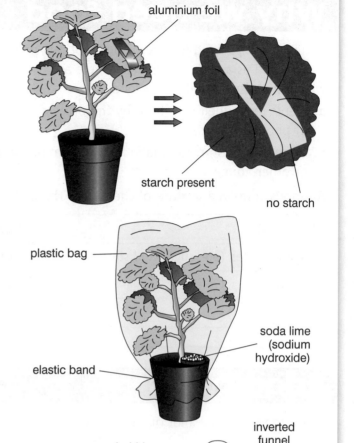

aluminium foil

starch present

no starch

Does a plant need carbon dioxide?

- De-starch a plant; enclose it in a clear bag containing sodium hydroxide (this absorbs carbon dioxide).

- After a few hours, test a leaf for starch; none should be present.

plastic bag

soda lime (sodium hydroxide)

elastic band

Is oxygen produced in photosynthesis?

- Set up the apparatus as shown in the diagram and collect the bubbles of gas.

- Test the gas for oxygen using a glowing splint. It should relight as the bubbles of gas were oxygen.

inverted funnel

oxygen bubbles

sunlight

weak hydrogen carbonate solution (gives CO$_2$ to plant)

plasticine supports

pond weed

Quick Test

1. How do you test for oxygen?

2. What is the chemical test for starch?

3. What colour does the leaf go if starch is present?

4. Why should you turn the Bunsen burner off when standing the test tube of ethanol in the hot water?

5. What three things does a plant need for photosynthesis?

6. What are the two things a plant produces during photosynthesis?

The need for food

Why we need food

All living things need food as a source of raw materials for growth and as a source of energy. There are two ways in which organisms obtain their food:

a. Plants make their own food by the process of photosynthesis.

b. Animals rely on ready-made food, by eating either plants or other animals.

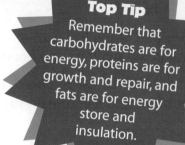

Top Tip
Remember that carbohydrates are for energy, proteins are for growth and repair, and fats are for energy store and insulation.

All foods contain a mixture of chemicals. The main chemicals are:

Chemical	Function
carbohydrates (e.g. sugar and starch)	provide energy
proteins	needed for growth and repair
fats	store energy, insulation, make cell membranes

Food types

These foods contain a lot of **carbohydrate**:

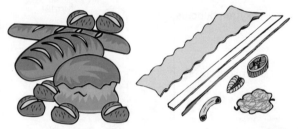

These foods contain a lot of **protein**:

These foods contain a lot of **fat**:

Chemical test for glucose

- Add a few drops of **Benedict's solution** to food solution.
- Heat in a water bath until it boils.
- If glucose is present, an **orange precipitate** will form.

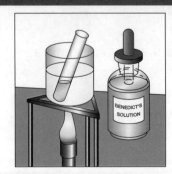

Chemical test for starch

- Add two drops of brown **iodine solution** to food solution.
- Solution will turn **blue/black** if starch is present.

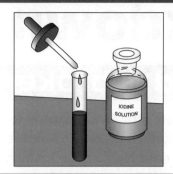

Top Tip
Learn the food tests for starch and glucose.

Chemical structure of food types

Carbohydrates, fats and proteins are mostly large molecules formed from many similar, smaller molecules linked together.

	Carbohydrate	Fat	Protein
Elements present	carbon (C) oxygen (O) hydrogen (H)	carbon oxygen hydrogen	carbon oxygen hydrogen nitrogen (N)
Basic units they are built from	glucose molecules	fatty acids and glycerol	amino acids
Diagram of structure	glucose molecules in a chain to form e.g. starch or cellulose	glycerol fatty acids	amino acids in a chain to form a protein

Quick Test

1. What do we use carbohydrates for?
2. Name the two main carbohydrates.
3. What is the chemical test for starch?
4. What is the chemical test for glucose?
5. What do our bodies need fat for?
6. Why is protein important to our cells?

7. Name the element found in proteins but not in fats or carbohydrates.
8. Name the basic units that proteins are built from.
9. Name the two basic units that fats are built from.

Mechanical breakdown of food

How food is broken down

Much of the food we eat must be changed before the body can use it. This involves the **breakdown of large insoluble food particles into smaller soluble particles** that can pass from the small intestine into the bloodstream.

The breakdown of food takes place in two steps:

a. the **mechanical breakdown** by **teeth**

b. the **chemical breakdown** by **enzymes**

e.g.   starch $\xrightarrow{\text{enzymes}}$ glucose
(insoluble)   (soluble)

First we will look at the mechanical breakdown of food – the first step in digestion.

Teeth

Human teeth

Humans have four kinds of teeth; each has a role in **breaking up food**.

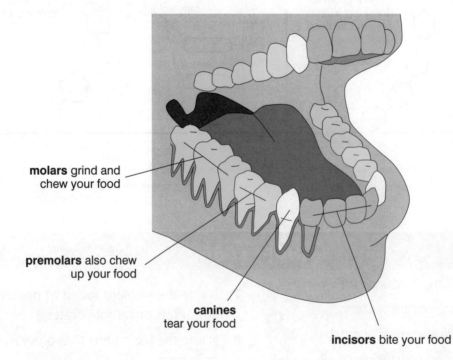

molars grind and chew your food

premolars also chew up your food

canines tear your food

incisors bite your food

You need to look after your teeth.

- Tooth decay is caused by **bacteria** in your mouth mixing with **saliva** to form **plaque**.
- Bacteria change the sugar in your food to **acid**, which attacks the enamel.
- Fluorides in toothpaste strengthen the enamel and make it more resistant to acid.

Animal dentition

Humans are **omnivores** – they can eat both meat and plants. Omnivores have teeth that are suitable for chewing both meat and plants.

Carnivores (meat eaters) and **herbivores** (plant eaters) have teeth to suit the food they eat.

Top Tip

Examine the dental arrangement of animals with different diets and note how their dentition is related to the type of food they eat.

Carnivore

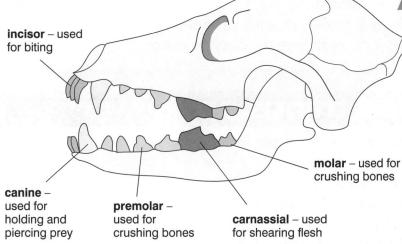

incisor – used for biting

canine – used for holding and piercing prey

premolar – used for crushing bones

carnassial – used for shearing flesh

molar – used for crushing bones

Herbivore

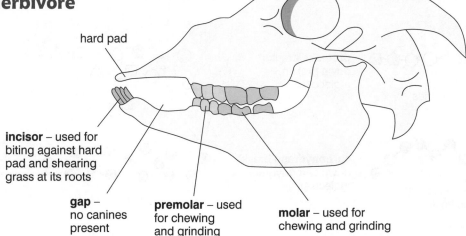

hard pad

incisor – used for biting against hard pad and shearing grass at its roots

gap – no canines present

premolar – used for chewing and grinding

molar – used for chewing and grinding

Quick Test

1. What is meant by the mechanical breakdown of food?

2. What is meant by the chemical breakdown of food?

3. Name the type of tooth used for biting.

4. What is the function of animal canine teeth?

5. Which type of teeth are used for chewing and grinding?

Answers: 1. The use of teeth to break down large lumps of food into smaller lumps. **2.** The action of enzymes which break down large insoluble molecules into soluble molecules. **3.** Incisors. **4.** Used for holding and piercing prey. **5.** Premolars and molars.

Chemical breakdown of food

Help with digestion

- Ultimately we need **nutrients** from our food to keep our bodies **healthy**.
- Remember: digestion breaks down large food molecules into small molecules so that they can pass into our bloodstream.
- After the **mechanical breakdown** of food by chewing, food must be broken down into smaller molecules as it passes through the digestive system.

Enzymes speed things up

- Carbohydrates, protein and fats are **large, insoluble food molecules**.
- Even after the teeth have done their bit and the stomach has churned the food up, the food molecules are still too big and insoluble to pass into the bloodstream.
- If you look at the diagram of the digestive system on page 38 you will see where the chemicals called enzymes are made.
- Enzymes are **specific**. There are **three main enzymes** in your system.

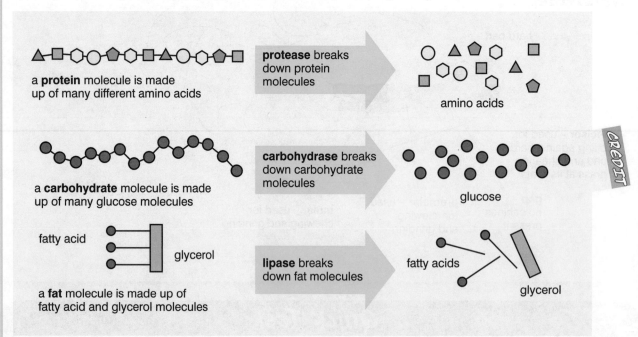

a **protein** molecule is made up of many different amino acids

protease breaks down protein molecules

amino acids

a **carbohydrate** molecule is made up of many glucose molecules

carbohydrase breaks down carbohydrate molecules

glucose

fatty acid

glycerol

a **fat** molecule is made up of fatty acid and glycerol molecules

lipase breaks down fat molecules

fatty acids

glycerol

CREDIT

- **Starch** is broken down into **glucose** in the **mouth** and **small intestine**.
- **Proteins** are broken down into **amino acids** in the **stomach** and the **small intestine**.
- **Fats** are broken down into **fatty acids** and **glycerol** in the **small intestine**.
- An example of a carbohydrase enzyme is **amylase**.
- An example of a protease enzyme is **pepsin**.

Experiment

- You can prove that a carbohydrase called amylase breaks down starch into sugar by setting up the following experiment.
- After leaving the test tube for 10 minutes in a water bath (maintained at 37ºC), test the water for starch (add iodine), and for glucose (add Benedict's solution and heat).
- Visking tubing acts like a model gut; it has tiny holes in it that will only allow small molecules through.

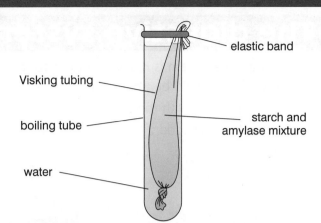

elastic band

Visking tubing

boiling tube

starch and amylase mixture

water

Results

- The water should test **negative for starch** (yellow) and **positive for sugar** (orange).
- The results show that the starch has been broken down into glucose.
- The glucose is small enough to pass through the visking tubing into the surrounding water.

Learn the names of the three main types of enzymes, where they are produced, what particular food type they act on, and what they break the food down into.

Quick Test

1. Name the type of enzyme that digests starch. Can you give an example?
2. Where are the enzymes that digest starch produced?
3. What does starch get digested into?
4. Name the type of enzyme that digests protein. Can you give an example?
5. Where is protein broken down?
6. What does protein get digested into?
7. Name the enzyme that digests fats.
8. What do fats get broken down into?

Answers 1. Carbohydrase, Amylase **2.** Mouth and small intestine **3.** Glucose **4.** Protease, Pepsin **5.** Stomach, small intestine **6.** Amino acids **7.** Lipase **8.** Fatty acids and glycerol

The digestive system

The digestive system explained

- The digestive system is really **one long tube called the gut**. If it were unravelled it would be **about nine metres long**!
- Digestion begins with the teeth and ends at the anus.
- It normally takes food 24–48 hours to pass through your digestive system.

Top Tip
Each part of the digestive system has a particular job. Learn the functions of each of the parts and where the enzymes and other helpful secretions are produced.

Mouth contains teeth that begin digestion by breaking up food

Oesophagus sometimes called the **gullet**

Liver

produces **bile**, which neutralises **stomach acid** and helps break down fat.

Salivary glands

secrete amylase which is a **carbohydrase enzyme** – **mucus** lubricates the food as it passes down the oesophagus.

Gall bladder

is where **bile** is stored until it is released into the small intestine via the bile duct.

Stomach

has muscular walls which churn up the food and mix it with the **gastric juices** that the stomach produces.
– The **gastric juices** contain **protease enzymes** and **hydrochloric acid**.
– The hydrochloric acid provides the acidic conditions for a protease enzyme called pepsin to work.

Pancreas

produces **carbohydrase, protease** and **lipase enzymes**.

Large intestine
receives any food that has not been absorbed into the blood. Excess water and salts are removed from the food. The remaining solid food turns into **faeces**.

Small intestine

also produces **all three** types of enzymes.
– This is where **digestion** is **completed** and dissolved food is **absorbed** into the **bloodstream**.
– The inner surface is covered in tiny finger-like projections called **villi**.

Rectum – where the faeces are stored before they leave the body via the **anus**.

Note: Food does not pass through the **pancreas**, **liver** and **gall bladder**. They are organs that **secrete enzymes** and bile to help digestion.

Absorption and the villus

- The small intestine is where the **digested food is absorbed into the blood**.
- The small intestine is well designed for absorption.
- It has a **thin lining**, a **good blood supply** and a **very large surface area**.
- The large surface area is provided by the **villi** (single = **villus**) that extend from the inside of the small intestine wall.

CREDIT

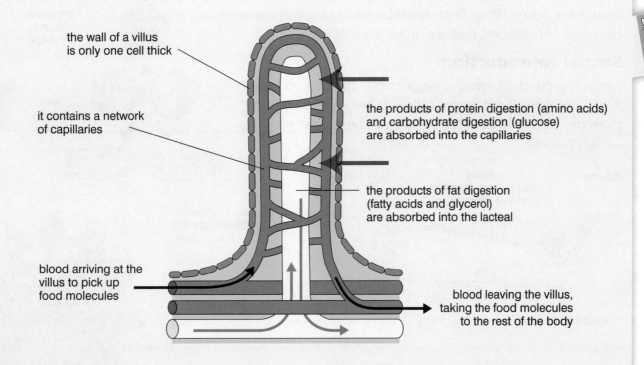

the wall of a villus is only one cell thick

it contains a network of capillaries

the products of protein digestion (amino acids) and carbohydrate digestion (glucose) are absorbed into the capillaries

the products of fat digestion (fatty acids and glycerol) are absorbed into the lacteal

blood arriving at the villus to pick up food molecules

blood leaving the villus, taking the food molecules to the rest of the body

Key Facts

- Digestion is the breaking down of large, insoluble molecules into small, soluble molecules so that they can be absorbed into the bloodstream through the small intestine wall.
- The large, insoluble molecules are starch (carbohydrates), protein and fat.
- This action is speeded up (catalysed) by enzymes.
- Enzymes in the small intestine are also found throughout the digestive system.
- Food is moved through the digestive system by the action of muscles in the gut wall. This is called **peristalsis**.

Quick Test

1. Where is the enzyme amylase produced?

2. Where in the digestive system does the food get absorbed into the bloodstream?

3. What enzymes does the pancreas produce?

4. Where are the villi found?

5. What happens in the large intestine?

6. Which organs produce enzymes?

7. Where is bile stored before it is released on to the food?

8. What are gastric juices?

Reproduction in animals

Types of reproduction

Reproduction can take place in two ways – **asexual** or **sexual**.

Asexual reproduction

Asexual reproduction only involves **one parent**. It does not involve the fusion of sex cells. **Amoeba** and **sea anenome** are examples of organisms that reproduce asexually.

Sexual reproduction

Sexual reproduction involves two parents. The parents have sex organs which produce sex cells. The sex cells are called **gametes**. The male gametes are **sperm**, the female gametes are **eggs** or ova.

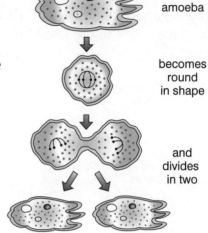

amoeba

becomes round in shape

and divides in two

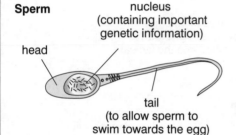

Sperm

head

nucleus (containing important genetic information)

tail (to allow sperm to swim towards the egg)

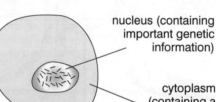

Egg (ovum)

nucleus (containing important genetic information)

cytoplasm (containing a large food store)

(not drawn to scale)

Top Tip
Make sure you understand the main differences between asexual and sexual reproduction.

Fertilisation and development

For sexual reproduction to take place a **sperm nucleus must fuse with an egg nucleus**. This is called fertilisation. There are two types of fertilisation:

a. **External fertilisation**: the eggs are released from the body into the external surroundings and the male then sheds his sperm over the eggs (e.g. fish).

b. **Internal fertilisation**: the eggs are fertilised by the sperm inside the female's body (e.g. mammals). This is important to land-living animals, as there is no water outside the body to carry the sperm to the egg.

Fertilisation

- In fertilisation a male gamete joins with a female gamete to produce a fertilised egg cell called a **zygote**.
- During fertilisation the 23 single chromosomes in the sperm cell pair up with the 23 chromosomes in the egg cell.
- Fertilisation restores the number of chromosomes to 46, or 23 pairs.
- During fertilisation it is a matter of chance which sperm fertilises which egg.
- **Sexual reproduction gives rise to variation** in the individual, as they will inherit a combination of the father's and mother's genes.

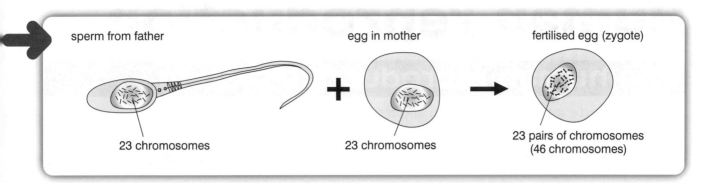

sperm from father   egg in mother   fertilised egg (zygote)

23 chromosomes   23 chromosomes   23 pairs of chromosomes (46 chromosomes)

External fertilisation

Trout development

Trout's eggs are laid in river beds; sperm are then released onto them which fertilises them. The young trout develops inside a **protective flexible covering**. After hatching, the young trout (**fry**) is able to look after itself. It feeds on **yolk** from the **yolk sac**. The trout fry begins to feed on **small water animals** and grows to adult.

Egg numbers and survival chances

Many sex cells and young are destroyed during the various stages of development by, for example:

a. eggs **not being fertilised**   **b.** eggs being **eaten**

c. eggs being **diseased**

The **greater the risks** involved in the type of reproduction, the **greater the number of eggs produced**.

Because the survival chances of the herring are very low (because **fertilisation** and **development** are **external**), the herring produces **very large numbers of eggs** to compensate and ensure that at least a small proportion will survive and become adults.

Animal	Number of eggs produced
herring	5,000,000
turtle	100

> **Top Tip**
> Make sure you understand the relationship between an organism's chances of survival and the number of eggs it produces.

CREDIT

Quick Test

1. Which type of reproduction involves only one parent?

2. Give an example of a single-celled organism that reproduces asexually.

3. Name the two types of gametes involved in sexual reproduction.

4. Name the type of fertilisation when the sperm reaches the egg outside the body of the female.

5. What type of fertilisation occurs in trout?

6. What does the young trout feed on after hatching?

7. List three ways sex cells and young are destroyed during the various stages of development of the trout.

8. If the survival chances of herring are low, how does this affect the number of eggs that they produce each year?

Answers 1. Asexual **2.** Amoeba, anemone **3.** Sperm and eggs (ova) **4.** External. **5.** External. **6.** Yolk from the yolk sac. **7.** Eggs not fertilised, eggs eaten by predators, eggs being diseased. **8.** Increased number of eggs to ensure that a small proportion survive to become adults.

Human reproduction

The human reproductive system

- From puberty, males produce **sperm** and females start to release **eggs**.
- Sperm is made in the **testes**.
- Eggs are produced in the **ovaries** and released into the **oviduct** (**fallopian tube**).

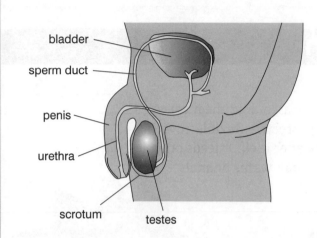

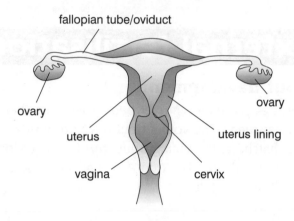

Internal fertilisation

- As part of the female menstrual cycle, an egg is released at around the middle of the cycle. This is called **ovulation**.
- When a man and a woman have **sexual intercourse**, sperm from the penis of the man passes into the vagina of the woman.
- The sperms swim up to the uterus and into the fallopian tubes (oviducts) to meet an egg.
- Many sperms die along the way. Only one sperm is able to break through the cell membrane of the egg and fertilise it.
- If a sperm meets an egg then fertilisation takes place.
- Fertilisation is the **fusing together** of the **sperm nucleus** and the **egg nucleus**.
- Fertilisation takes place in the **fallopian tube** / oviduct.

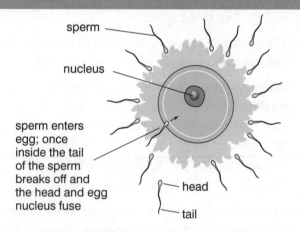

After fertilisation

- The fertilised egg divides into a **ball of cells** as it passes down the fallopian tube.
- The ball of cells becomes an **embryo** and embeds itself into the uterus lining. This is called **implantation**.
- The embryo develops into a baby.
- At about nine weeks the embryo is now called a **foetus**.

Development and protection

- During development the foetus is provided with food and oxygen by the **umbilical cord**.
- Waste materials from the foetus pass back along the umbilical cord.
- The blood of the foetus and the mother do not mix but they pass close together to allow **exchange of food**, **oxygen** and **wastes**.
- The **placenta** is an organ that grows early in the pregnancy. It acts as a **barrier**, **preventing harmful substances** reaching the foetus.
- The foetus is attached to the placenta by the umbilical cord.
- The baby is **protected** inside the uterus by the amniotic sac filled with a watery liquid, called the **amniotic fluid**.
- The fluid acts as a **shock absorber** against minor bumps.

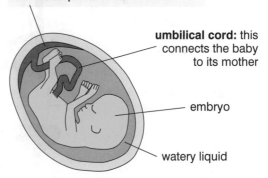

placenta: here the embryo's vessels are close to the mother's, and food and oxygen diffuse into the embryo and waste products diffuse out

umbilical cord: this connects the baby to its mother

embryo

watery liquid

Birth and after

- After nine months of pregnancy the baby is ready to be born through the **vagina**.
- The baby normally turns so that its head is down towards the **cervix**.
- Muscles in the wall of the uterus begin to **contract** and the cervix widens.
- The baby's head passes through the cervix when it has dilated enough, into the vagina.
- The **fluid sac bursts** and the watery liquid runs out.
- **Contractions** push the baby out of the vagina.
- More contractions push the placenta out. This is called the **afterbirth**.
- The umbilical cord is cut and the baby has to breathe for itself.
- After birth, human babies (and all mammal young) are dependent on the adult for **care** and **protection**.

Top Tip
Learn the stages leading up to implantation of the embryo: ovulation, sexual intercourse, fertilisation, cell division and implantation.

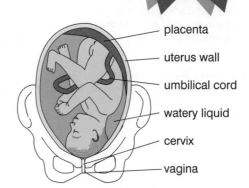

placenta

uterus wall

umbilical cord

watery liquid

cervix

vagina

Quick Test

1. What does a fertilised egg develop into?
2. What is implantation?
3. Where does fertilisation take place?
4. What is fertilisation?
5. How is the foetus supplied with oxygen and food while in the uterus?
6. How long does a human pregnancy usually last?

Answers 1. An embryo **2.** The embryo embedding itself into the uterus lining. **3.** In the fallopian tube **4.** The fusing together of the sperm nucleus with the egg nucleus **5.** By the umbilical cord **6.** Nine months/40 weeks

Water and waste

The kidneys and waste

The kidneys are vital because they **filter the blood** of poisonous waste and **regulate** the amount of water in the body.

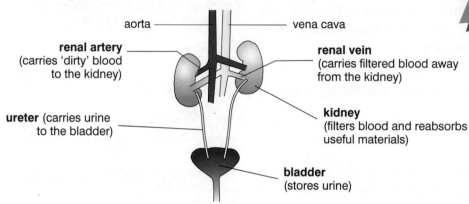

aorta — — vena cava

renal artery
(carries 'dirty' blood to the kidney)

renal vein
(carries filtered blood away from the kidney)

ureter (carries urine to the bladder)

kidney
(filters blood and reabsorbs useful materials)

bladder
(stores urine)

The kidneys remove the poisonous substance **urea**.

- All your blood is filtered by the kidneys about 300 times per day.
- The filtering is done by millions of tiny little tubes called **nephrons**.

A nephron

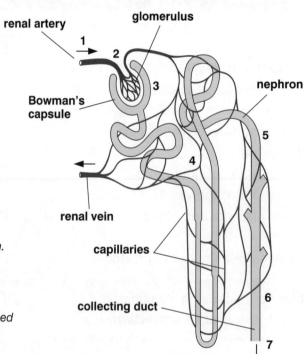

renal artery

glomerulus

1

2

Bowman's capsule

3

nephron

5

4

renal vein

capillaries

collecting duct

6

7

1 *Unfiltered blood enters the nephron along a branch of the renal artery.*

2 *The renal artery divides into a group of capillaries called the glomerulus.*

3 *Blood is filtered through the glomerulus and Bowman's capsule and into the rest of the nephron.*

4 *As the liquid passes along the nephron, useful substances (like glucose and vitamins) are reabsorbed back into the blood, which passes out of the kidney along the renal vein.*

5 *The liquid left in the nephron contains urea, water and other unwanted substances.*

6 *The waste liquid passes into the collecting duct.*

7 *The collecting duct empties into the bladder.*

How urea is produced

During digestion, enzymes break **proteins** down into **amino acids**, which are then built into new proteins by the **liver**. Any amino acids not used are broken down by the liver into **urea**. The urea is then filtered out by the kidneys.

The kidneys and water balance

Our bodies are 60% water. Too much or too little water in a cell affects its functions and is dangerous.

You can gain water or lose water in different ways, but the volume of water we gain must equal the volume of water we lose. This is called **water balance**. The kidneys help to maintain this balance by producing more or less urine depending on the body's water intake.

How water enters and leaves the body			
daily water gain	cm³	daily water loss	cm³
food and drink	1400	urine	700
water made in **respiration** (when energy is released from food)	350	sweat	500
		breathing	400
		faeces (solid waste)	150
Total	1750	Total	1750

Water balance

Water balance is controlled by a chemical produced in the brain called **antidiuretic hormone** (ADH).

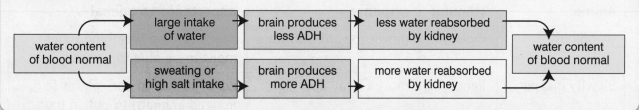

CREDIT

Kidney damage

Kidney failure is fatal unless it is treated using a **kidney machine** or by a **kidney transplant**. Kidney machines are very expensive and patients must spend a lot of time connected to them. Kidney transplants are not frequent because of a shortage of donors. The transplanted kidney can also be rejected by the body of the patient.

Quick Test

1. Give two reasons why the kidneys are important.
2. Name the blood vessel that takes blood towards the kidney.
3. Name the tube that carries urine from the kidney to the bladder.
4. List three ways in which the body gains water.
5. List four ways in which the body loses water.
6. Name the waste that is filtered out by the kidneys.
7. What happens to useful substances that have been filtered out into Bowman's capsule?
8. Name the hormone that controls the amount of water reabsorbed in the kidneys.
9. Give two ways in which people with kidney failure are treated.

Answers 1. They help get rid of poisonous waste; they regulate the amount of water in the body. **2.** Renal artery **3.** Ureter **4.** Drinking, in food, chemical reactions (eg respiration) **5.** In urine, sweat, breathing out, faeces **6.** Urea **7.** They are selectively reabsorbed (eg glucose and vitamins) **8.** ADH (anti-diuretic hormone) **9.** Kidney machines; kidney transplants

Responding to the environment

Key Facts

Environment and survival
- Living things are found on land, in the air and in water. These surroundings are their **environment**.
- An environment has many places, called **habitats**, where animals or plants can live.
- Each organism has its own habitat with the **conditions** it needs to **survive**.

Responses to changes in the environment

The way in which an animal **responds** to a **stimulus** from its surroundings is important to the animal's survival. The conditions an animal chooses are the best for its survival.

Animal	stimulus	response	importance to animal
Woodlouse	moisture	moves towards moisture	needs to keep breathing system moist to be able to breathe
Blowfly maggot	light	moves away from light	needs to stay in dark places (e.g. in dead animals) to obtain food and protection from predators

Rhythmical behaviour

Some environmental conditions change regularly. These changes can be **daily** changes (such as night to day) or **seasonal** changes (such as hours of daylight and temperature).

Many animals change their behaviour in response to these regular changes in the environment. These changes are examples of **rhythmical behaviour**.

The stimulus that sets of the change is called a **trigger stimulus**.

Animal	rhythmical behaviour pattern	trigger stimulus	importance to animal
Hedgehog	hibernation	decreasing day length and temperature	conserves energy when food is in short supply
Cockroach	night activity	darkness	less chance of being caught by predators if they feed at night

Decreasing day length triggers some birds to migrate.

Learn other examples of rhythmical behaviour (e.g. migration) in other animals and the stimuli that trigger these behaviours.

Quick Test

1. Why is it important that animals respond to changes in their environment?

2. Why do blowfly maggots move away from the light?

3. Explain why the response of woodlice to environmental stimuli increases their chances of survival.

4. What is rhythmical behaviour?

5. What is meant by trigger stimulus?

6. Give an example of rhythmical behaviour shown by hedgehogs.

7. What trigger stimuli cause this behaviour in hedgehogs?

8. Complete the table below which summarises rythmical behavious.

Animal	Rhythmical behaviours	Trigger stimulus
Swallow	a	Decreasing day length
b	Hibernation	c

Answers 1 The conditions an animal chooses are usually the best for its survival. 2 By staying in dark places they are more likely to obtain food and protection. 3 They move to where it is damp where they can breathe easier. 4 Behaviour of animals in response to regular changes in the environment. (eg temperature, day length). 5 An environmental condition or change that sets off a behavioural response. 6 Hibernation. 7 Decreasing day length and temperature. 8 a Migration b Hedgehog etc c Decreasing temperature

Cells and diffusion

- Cells are the building blocks of life.
- All living things are made up of cells.
- A living thing is called an organism.
- Plants and animals are organisms.

Animal cell

Each part of the cell has a special job to do.

Cell membrane – controls what passes in and out of the cell.

Nucleus – controls all the chemical reactions that take place inside the cell. The nucleus also contains chromosomes that contain **all the information** needed to produce a new living organism.

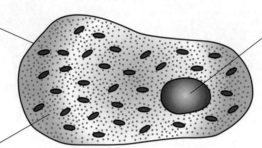

Cytoplasm – where all the **chemical reactions** take place.

Top Tip
Stains are used in the preparation of microscope slides to make cell structures (e.g. nucleus, cytoplasm, membrane) more clearly visible.

Plant cell

Each part of the cell has a special job to do.

Chloroplasts – contain a green substance called **chlorophyll**. This absorbs the sun's energy so that the plant can **make its own food** during **photosynthesis**.

Nucleus

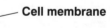

Cell membrane

Cytoplasm

Cell wall – made of cellulose, which gives a plant cell **strength and support**.

Vacuole – contains a weak solution of salt and sugar called **cell sap**. The vacuole also gives the cell support.

Similarities and differences

The typical plant cell and animal cell look different to each other, for example, they are different shapes and the **nucleus** is in a different position.

They both have:	Only plant cells have:
Nucleus	Cell wall
Cytoplasm	Vacuole
Cell membrane	Chloroplasts

Top Tip
Make sure you know the similarities, and in particular the differences, between an animal and a plant cell.

Simple diffusion

Diffusion is **the movement of molecules from an area of high concentration to an area of low concentration**.

In other words, it is the natural tendency of molecules to move into all the available space until they are **evenly spread out**.

Two rules to remember:

- The **larger** the molecule, the **slower** the rate of diffusion.
- **The greater the difference in concentration, the greater the rate of diffusion**. The difference is called a **concentration gradient**.

Example of diffusion in animal cells

- Body cells need food and oxygen for **respiration**.
- These are carried in the blood.
- When the blood reaches the cells, the oxygen and food **diffuse** into the cells.
- Your cells produce waste and carbon dioxide.
- These **diffuse** out of the cells into the blood.
- The exchange of carbon dioxide and oxygen between the alveoli in the lungs and the blood is another example of diffusion.

Example of diffusion in plant cells

- A plant needs carbon dioxide for **photosynthesis**.
- Carbon dioxide **diffuses** into a leaf via the stomata (holes) found on the underside of a leaf.
- A leaf produces oxygen and water vapour.
- Oxygen and water vapour **diffuse** out of the stomata.
- Diffusion of water vapour occurs much quicker in **hot**, **dry**, **windy** conditions. Think about the best conditions for drying clothes on a line.

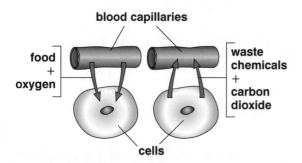

blood capillaries

food + oxygen

waste chemicals + carbon dioxide

cells

leaf cells

stoma

carbon dioxide   water & oxygen

Quick Test

1. Name three differences between an animal and a plant cell.

2. Name three similarities between an animal and a plant cell.

3. What does the cell membrane do?

4. What does the cell wall do?

5. What substance do chloroplasts contain and what does it do?

6. What is the job of the nucleus?

7. What is diffusion?

8. Give an example of diffusion in plant cells.

9. Give an example of where diffusion takes place in an animal.

10. The greater the concentration difference the _____ the diffusion rate?

Osmosis

Key Facts

Each cell is surrounded by a cell membrane which has tiny holes in it. This membrane is **partially permeable**. It allows small molecules to pass through, but not larger ones.

Osmosis – a special case of diffusion

The definition of osmosis

The movement of water molecules from an area of high water concentration (weak solution) **to an area of low water concentration** (strong solution) **through a partially permeable membrane**.

Water moves both ways to balance up the concentrations. If there is more movement one way, we say there is a **net movement** of water into the area where there is less water.

Top Tip
Learn the definitions of diffusion and osmosis.

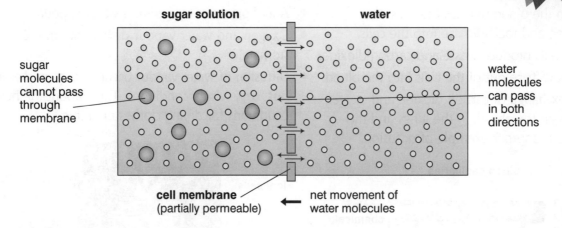

sugar solution   water

sugar molecules cannot pass through membrane

water molecules can pass in both directions

cell membrane (partially permeable)   net movement of water molecules

Osmosis makes plant cells swell. Water moves into the plant cell vacuole and pushes against the cell wall. The cell wall stops the cell from bursting. We say that the cell is **turgid**. This is useful as it gives plant stems support.

If a plant lacks water, it wilts and the cells become **flaccid** as water has moved out of the cell.

If a lot of water leaves the cell, the cytoplasm starts to peel away from the cell wall. We say the cell has undergone **plasmolysis**.

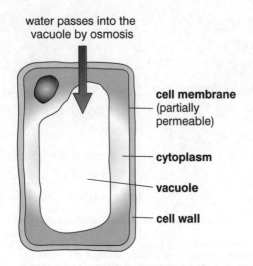

water passes into the vacuole by osmosis

cell membrane (partially permeable)

cytoplasm

vacuole

cell wall

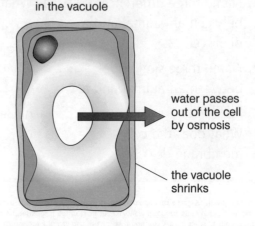

the solution outside the cell is more concentrated than in the vacuole

water passes out of the cell by osmosis

the vacuole shrinks

Osmosis in plant cells

Root hairs take water from the soil by **osmosis**. Water continues to move along the cells of the root and up the **xylem** to the leaf. Water always moves to areas of lower water concentration.

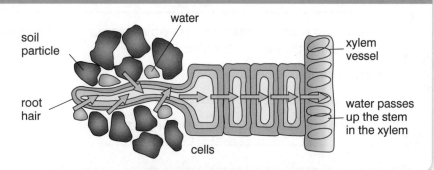

Osmosis in animal cells

Animal cells have no cell wall to stop them swelling. So if they are placed in pure water, they take in water by osmosis until they burst.

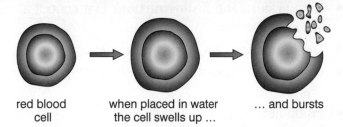

red blood cell   when placed in water the cell swells up …   … and bursts

An important osmosis experiment

Visking tubing is a **partially permeable membrane** which allows water in and out, but not sugar molecules. It is used to show the effects of **osmosis**.

- The visking tubes are filled with a weak sugar solution.
- In pure water the visking tubing swells and becomes **turgid** as water enters by **osmosis**.
- In strong sugar solution water leaves the visking tubing by **osmosis** and it becomes **flaccid**.
- The greater the concentration gradient, the faster the rate of osmosis.

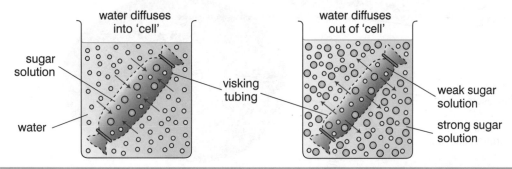

Quick Test

1. What is a partially permeable membrane?

2. What substance moves by osmosis?

3. What happens to plant cells that take up water by osmosis?

4. What happens to plant cells that lose water by osmosis?

Cell division

Cell division is an important process which **increases** the number of **cells** in an organism. The division of the nucleus is called **mitosis**.

Mitosis is important because:

a. it produces new cells for **growth** and **repair**

b. it is the way in which single-celled organisms **reproduce**.

Nuclei, chromosomes and genes

- Inside nearly all cells is a **nucleus**.
- The nucleus **contains instructions** that control all cell activities, including mitosis.
- The instructions are carried on **chromosomes**.
- **Genes** on the chromosome control each particular characteristic.
- Inside human cells there are **23 pairs** or **46 single chromosomes**. The cell is called **a diploid cell**.
- Other animals have different numbers of chromosomes.

Top Tip
It is crucial to understand where the chromosomes and the genes originate from in order to understand how information is passed on from one generation to the next.

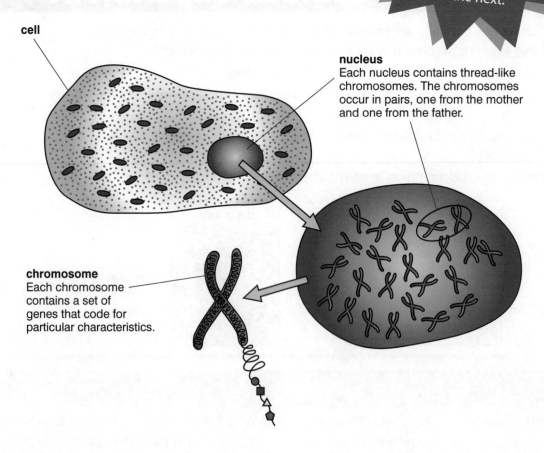

cell

nucleus
Each nucleus contains thread-like chromosomes. The chromosomes occur in pairs, one from the mother and one from the father.

chromosome
Each chromosome contains a set of genes that code for particular characteristics.

Mitosis

Summary of mitosis and cell division

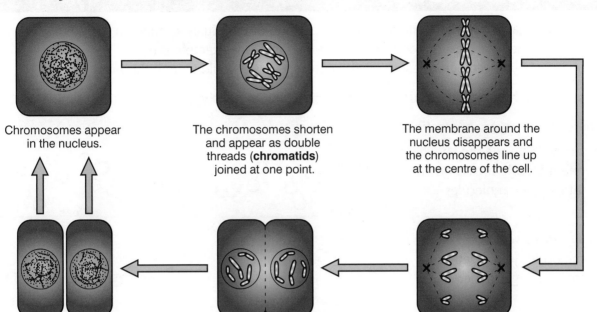

Chromosomes appear in the nucleus.

The chromosomes shorten and appear as double threads (**chromatids**) joined at one point.

The membrane around the nucleus disappears and the chromosomes line up at the centre of the cell.

The two new cells now go through a period of growth before mitosis starts again in each cell.

The nucleus membrane reforms round each group of chromatids and the cytoplasm divides.

The chromatids are pulled apart and move to opposite ends of the cell.

Why must a new set of instructions be made?

It is important that the **full set of chromosomes is passed to each of the daughter cells**.

Occasionally, an extra chromosome goes into one cell, meaning one daughter cell has a chromosome missing, and one has an extra chromosome. This means the cells have some extra or missing information. This is called **chromosome mutation** and can have serious health effects on offspring. See the 'Inheritance' chapter for more details.

Top Tip
Make sure you can put diagrams of mitosis into the right sequence.

Quick Test

1. What are the thread-like structures contained in the nucleus called?

2. How many chromosomes do humans have in their body cells?

3. What are genes?

4. What name is given to the cell division process that takes place during growth and development?

5. As a result of mitosis, the chromosome numbers in the daughter cells are the same as in the parent cell. Is this statement true or false?

Answers 1. Chromosomes. 2. 46 3. Sections of chromosomes that code for a particular characteristic 4. Mitosis 5. True

Enzymes

Cell reactions

A chemical reaction is where one molecule, or a group of molecules, changes in some way. Chemical reactions occur in cells all the time. One example is the changing of hydrogen peroxide into water and oxygen. We would write it this way:

hydrogen peroxide ⟶ **water + oxygen**

There are two types of reaction:

a. **breakdown reactions:** where one molecule breaks down into smaller molecules.

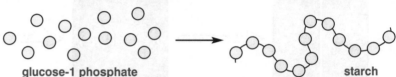

starch ⟶ maltose

b. **synthesis reactions:** where molecules join together to make a larger molecule.

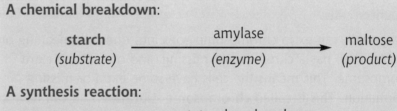

glucose-1 phosphate ⟶ starch

What are enzymes?

Catalysts are chemicals which speed up reactions. **Enzymes are biological catalysts** which speed up reactions that take place in living cells, All cell reactions are speeded up, or helped, by enzymes.

The molecule that an enzyme works on is called the **substrate**.

The molecule that is produced is called the **product**.

A chemical breakdown:

$$\text{starch} \xrightarrow{\text{amylase}} \text{maltose}$$

(substrate)   (enzyme)   (product)

A synthesis reaction:

$$\text{glucose-1-phosphate} \xrightarrow{\text{potato phosphorylase}} \text{starch}$$

(substrate)   (enzyme)   (product)

More information about enzymes

a. Enzymes are made from **protein**.

b. They can be used **over and over again**. They are **never used up** in reactions.

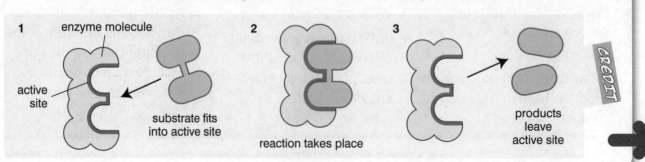

1   enzyme molecule

active site

substrate fits into active site

2   reaction takes place

3   products leave active site

CREDIT

c. They are **specific**. This means that **each enzyme can only react with one set of substrate molecules**. For example, the enzyme catalase will only affect the breakdown of hydrogen peroxide into water and oxygen. It does not affect any other reaction.

CREDIT

d. Enzymes are affected by temperature:

(i) They work faster at warm temperatures.

(ii) At high temperatures (over 50°C) many enzymes are **denatured**. This means that they are altered in shape and cannot work.

(iii) The temperature they work **best** at is called the **optimum** temperature. The optimum temperature for most mammal enzymes is 37°C. Some plant enzymes have an optimum temperature around 20°C.

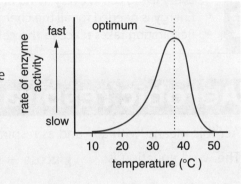

e. Enzymes are affected by pH. The pH an enzyme works best at is called its optimum pH.

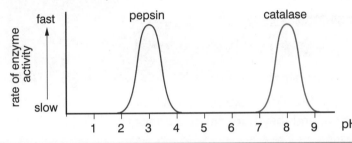

Pepsin works best between pH 2 and pH 4. Its optimum is pH 3.

Catalase works best between pH 7 and pH 9. Its optimum is pH 8.

Enzymes and cell chemistry

Almost all the reactions that happen inside an organism are controlled by enzymes. Without enzymes these reactions would go too slowly for life to exist. Enzymes are essential for life. Most processes in an organism consist of many reactions, each of which is catalysed by an enzyme. These processes are what we call **metabolism**. Some examples are given below:

a. photosynthesis   **b.** respiration   **c.** germination   **d.** growth   **e.** digestion

Top Tip
Learn about all the factors that affect the rate of enzyme reactions as they are often asked for in examination questions.

Quick Test

1. What are enzymes?

2. Name two types of enzyme reaction.

3. Give an example of a reaction in which one molecule is broken down into smaller molecules.

4. What type of chemical molecule are enzymes made of?

5. What word describes an enzyme's ability to only catalyse one type of reaction?

6. Name two environmental factors that affect the rate of enzyme reactions.

7. What term is used to describe the temperature at which an enzyme works best?

8. Name two processes in organisms that depend on enzymes.

Answers 1. Biological catalysts that speed up chemical reactions. **2.** Breakdown and synthesis. **3.** Starch into maltose. **4.** Protein. **5.** Specific. **6.** Temperature, pH. **7.** Optimum. **8.** Photosynthesis, respiration, germination, growth, digestion.

Aerobic respiration

- Respiration is **not** breathing or the exchange of gases.
- **Respiration is the breakdown of glucose to release energy using oxygen.**
- Every living cell in every living organism uses respiration to release energy, all of the time.
- Energy is needed for all the chemical reactions in the body.
- Respiration takes place **in the cell cytoplasm**.

Aerobic respiration

Aerobic means 'with air' and as respiration needs oxygen, we call it **aerobic respiration**.

The word equation is: **glucose + oxygen ⟶ carbon dioxide + water + energy**

The **carbon dioxide** and **water** are **waste products**, removed from the body via the lungs, skin and kidneys.

Top Tip
Learn the word equation for respiration.

Similarities between respiration and burning

Respiration inside every living cell **is similar to burning** – except there are no flames. They **both require a fuel and oxygen**, and they **both release energy and waste** products.

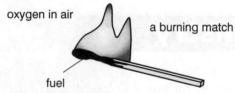

oxygen in air
a burning match
fuel

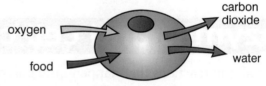

oxygen
food
carbon dioxide
water

fuel (wood) + oxygen
→ carbon dioxide + water + energy

food (glucose) + oxygen
→ carbon dioxide + water + energy

Measuring the energy content in food

- By burning a sample of food, we can see how much energy it releases in heating up a volume of water.
- The temperature of the water is taken at the beginning.
- The food substance is set alight and held underneath the water.
- When the food has stopped burning the temperature of the water is taken again. The difference in temperature is recorded.

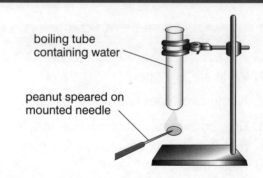

boiling tube
containing water

peanut speared on
mounted needle

All foods contain energy (a **kilojoule (kJ)** is a unit of energy), but some contain more than others. Fats and oils contain twice as much energy as proteins and carbohydrates.

fats and oils	38 kJ per gram
carbohydrates	17 kJ per gram
proteins	22 kJ per gram

Uses of the energy

The chemical energy in food is converted during respiration and used for:

- **making your muscles work** (your muscles need a lot of energy during exercise)
- **chemical reactions**
- **growth** and **repair** of cells
- **making up larger molecules** from smaller ones, i.e. proteins from amino acids
- **maintaining body temperature** in warm-blooded animals.

Top Tip
Make sure that you understand the difference between respiration and breathing.

Measuring respiration

During **aerobic respiration**, **oxygen** gas is used up and **carbon dioxide** gas is produced.

A **respirometer** can be used to show these gas changes during respiration.

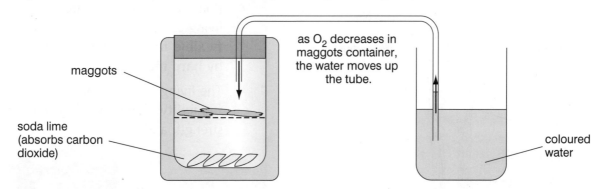

maggots

soda lime (absorbs carbon dioxide)

as O₂ decreases in maggots container, the water moves up the tube.

coloured water

As the oxygen is used up by the maggots, the coloured water moves towards them up the glass tube. The carbon dioxide they breathe out is absorbed by the soda lime and so does not push the water back.

We can place the maggots in different conditions (e.g. hot or cold) and compare the rate of movement of the coloured water. This is a measure of the rate of respiration.

Quick Test

1. What is aerobic respiration?

2. Where does respiration take place?

3. Where does the oxygen come from that enables animals to carry our respiration?

4. What are the waste products of respiration?

5. How much energy is stored in fats compared to carbohydrates and proteins?

6. List three ways in which energy released during respiration is used by organisms.

7. What energy change takes place so that energy is released to keep you warm?

8. Name the piece of equipment that is used to measure the rate of respiration.

Answers 1. Breaking down glucose in the presence of oxygen to release energy. **2.** In the cytoplasm of cells. **3.** Breathed in from the atmosphere. **4.** Water and carbon dioxide. **5.** About twice the amount in fats compared to carbohydrates and protein. **6.** Movement, growth, repair, body heat, chemical reactions. **7.** Chemical energy in food ← heat energy. **8.** Respirometer

Movement

The human skeleton

The skeleton has a number of functions:

a. it keeps our **shape**

b. it **supports** our weight

c. it **protects vital organs** such as the brain, heart, lungs and spinal cord

d. it provides **a framework for** the attachment of **muscles**

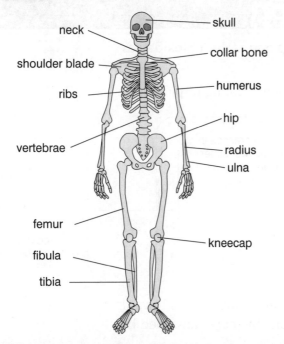

- neck
- skull
- shoulder blade
- collar bone
- ribs
- humerus
- vertebrae
- hip
- radius
- ulna
- femur
- kneecap
- fibula
- tibia

The structure of bone

Bones are alive. They are made of living cells. These cells are made of **flexible fibres** and are surrounded by **hard minerals**.

The living cells give bones **flexibility**. The minerals give bones strength and **hardness**.

Top Tip
Remember tendons join muscle to bone and ligaments join bone to bone.

Joints

A **joint** is a place where two or more bones meet (e.g. elbow or shoulder).

Different joints in the body allow different types of movement.

The bones are held together by strong fibres called **ligaments**.

Ball and socket and **hinge** joints are also known as **synovial joints**.

Type of joint	Types of movement possible	Examples
ball and socket	in all directions	shoulder and hip
hinge joint	in one plane (like a hinge)	knee and elbow

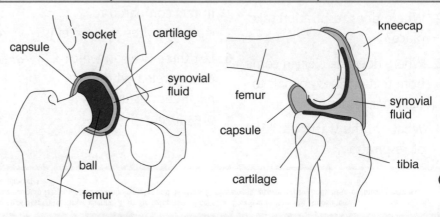

- capsule
- socket
- cartilage
- synovial fluid
- ball
- femur
- kneecap
- femur
- synovial fluid
- capsule
- cartilage
- tibia

The ends of the bones in synovial joints have a layer of **smooth cartilage**.

Cartilage acts as a **shock absorber** that prevents the wearing away of surfaces. The cartilage is covered by **synovial fluid**. **Synovial fluid** helps **reduce friction** at the joint.

Top Tip
Questions about the structure and location of synovial joints are often asked in examinations.

Muscles and movement

- The **muscles** provide the force needed to move the bones at joints.
- Muscles can only pull; they cannot push.
- When a muscle pulls it gets shorter and fatter: it **contracts**.
- When a muscle is not contracting it **relaxes** and returns to its normal size.
- Muscles all over the body **work in pairs**; while one contracts the other relaxes.
- These are called **antagonistic pairs** because they work in opposite directions to produce movement.
- Muscles are attached to bones by **tendons**.
- Tendons are **very strong** and **do not stretch** so muscles can pull on the bone effectively.

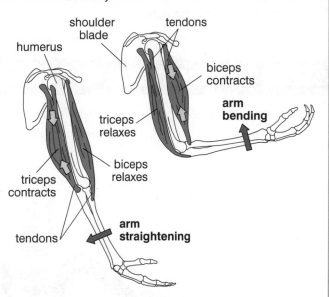

Sports injuries

If you get involved in a strenuous activity such as sport, it is possible to injure your joints, muscles and tendons. Some parts of the body are more likely to be injured than others. The most common injury is to the knee and the least common is to the head and neck. Sports injuries occur because of the sudden changes of movement and the knocks that can happen.

Quick Test

1. List the functions of the skeleton.
2. Name the two main components of bone.
3. Where in the body are ball and socket joints found?
4. Where in the body are hinge joints found?
5. Give two functions of cartilage.
6. What is the function of ligaments?
7. What is the function of synovial fluid?
8. Name the end part of a muscle that attaches it to a bone.
9. What term is used to describe pairs of muscles in a limb that act in opposition to one another?
10. Which part of the body is most commonly injured during sporting activities?

Answers 1. Keeps the shape of the animal, support, protection, framework for muscle attachment **2.** Flexible fibres, minerals **3.** Shoulder and hip **4.** Elbow, knee, fingers **5.** Acts as shock absorber, provides tough smooth surface where bones meet **6.** Hold bones together at a joint **7.** It acts as a lubricant in the joint **8.** Tendon **9.** Antagonistic pairs **10.** Knee

Energy and the breathing system

Breathing for energy

To release the energy in food, it must be combined with oxygen in the body during a process called respiration.

Breathing supplies the cells with **oxygen** and removes the waste product of respiration, **carbon dioxide**.

The structure of lungs

The oxygen you need to release energy from food is obtained by breathing in. The carbon dioxide produced as a waste product is removed by breathing out.

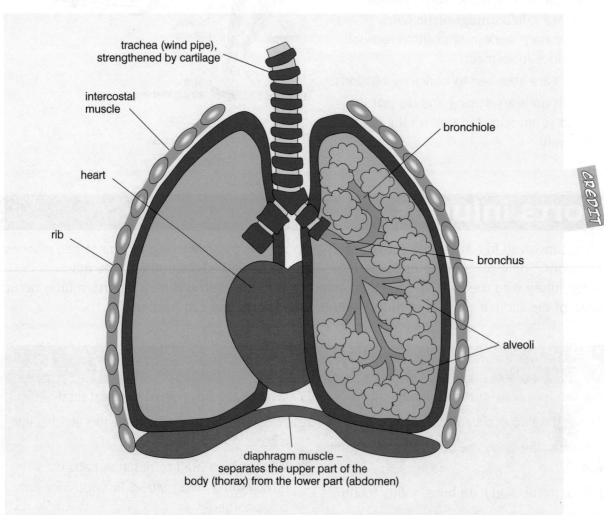

trachea (wind pipe), strengthened by cartilage

intercostal muscle

heart

rib

bronchiole

bronchus

alveoli

CREDIT

diaphragm muscle – separates the upper part of the body (thorax) from the lower part (abdomen)

Air comes in at the mouth and passes down the **trachea**. On the way, it passes through the larynx (voice box). The trachea is held open by **rings of cartilage**. The air then passes down a **bronchus**, which divides into many tiny tubes called **bronchioles**. The bronchioles end in **air sacs** – alveoli.

Gas exchange in the air sacs

- The alveoli are well designed for their job of gas exchange.
- There are millions of alveoli, so they present a **large surface area**; they are in **very close contact** with lots of blood capillaries.
- Their surface lining is **moist**, so that the gases can dissolve before they diffuse across the **thin membrane**.
- At the lungs, oxygen diffuses into the blood and carbon dioxide diffuses into the alveoli.
- At the cells, oxygen diffuses into them and carbon dioxide diffuses out into the blood.

CREDIT

Top Tip
Credit students should know the characteristics of a good absorbing surface, as this is often asked in the exam.

Top Tip
Make sure you know the structure of the breathing system.

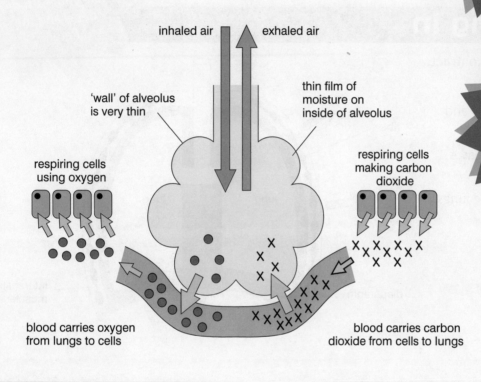

inhaled air   exhaled air

'wall' of alveolus is very thin

thin film of moisture on inside of alveolus

respiring cells using oxygen

respiring cells making carbon dioxide

blood carries oxygen from lungs to cells

blood carries carbon dioxide from cells to lungs

Quick Test

1. Where does gas exchange take place?
2. Why are the alveoli so good at gas exchange?
3. Why is breathing important?
4. What do the bronchi branch up into?
5. What strengthens the trachea?
6. What is the upper part of the body called?
7. What is the lower part of the body called?
8. What separates the upper body from the lower body?

Answers 1. The alveoli **2.** Large surface area, thin walls, moist and close to blood capillaries **3.** For respiration **4.** Bronchioles **5.** Cartilage **6.** Thorax **7.** Abdomen **8.** Diaphragm

More about the breathing system

Key Facts

Breathing
- **Breathing** is a physical process that **delivers oxygen** to the alveoli in your lungs and **removes carbon dioxide.**
- Breathing is made possible by the action of two muscles, the **diaphragm** and the **intercostal muscles.**

Top Tip
The diagrams of breathing out and in look similar so make sure you learn the detail and the differences.

Breathing in

Intercostal muscles **contract**.

Ribs move **up** and **out**.

Diaphragm **contracts** and **flattens**.

Thorax volume **increases**, pressure **decreases**.

Air is drawn **into** the lungs.

CREDIT

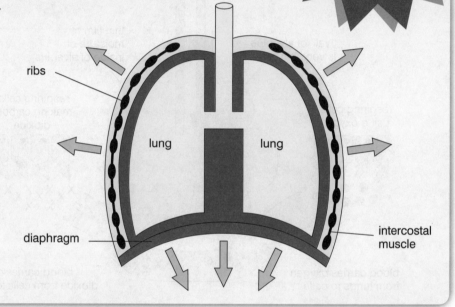

ribs

lung   lung

diaphragm

intercostal muscle

Breathing out

Intercostal muscles **relax**.

Ribs move **down** and **in**.

Diaphragm **relaxes** and **moves up**.

Thorax volume **decreases**, pressure increases.

Air is forced **out of** the lungs.

The movement of air into and out of the lungs is called **ventilation**.

CREDIT

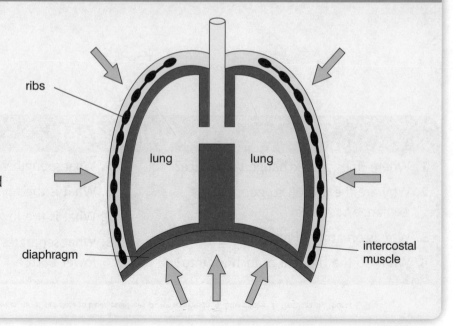

ribs

lung   lung

diaphragm

intercostal muscle

Composition of gases

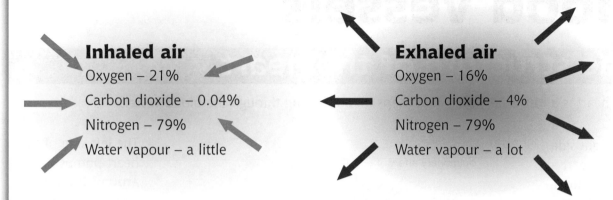

Inhaled air
Oxygen – 21%
Carbon dioxide – 0.04%
Nitrogen – 79%
Water vapour – a little

Exhaled air
Oxygen – 16%
Carbon dioxide – 4%
Nitrogen – 79%
Water vapour – a lot

Note

- Notice that we breathe out oxygen and carbon dioxide as well as breathing them both in.
- It is important to note that we breathe in **more** oxygen and breathe out **more** carbon dioxide.
- You should also notice that the amount of water vapour is higher in the air that we breathe out. There are other important differences described above.

Ciliated epithelial cells

The ciliated cells are specialised cells that line your nose and lung passages. These cells make a slimy liquid called **mucus**.

- We warm, moisten, filter and clean the air we breathe in.
- Dust and germs get trapped in the mucus. Tiny hairs called **cilia** move the mucus to the nose and throat to prevent germs reaching the lungs.
- The air we breathe out is actually cleaner.

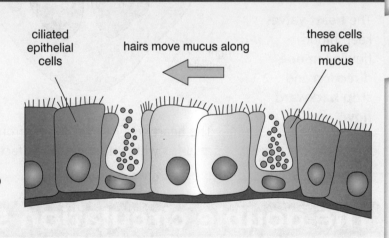

ciliated epithelial cells

hairs move mucus along

these cells make mucus

CREDIT

Quick Test

1. What happens to the ribs when you breathe out?
2. What happens to the diaphragm when you breathe in?
3. What do ciliated cells produce that trap dirt and bacteria?
4. What are cilia?
5. Give four differences between the air we breath out and the air we breath in.

Answers 1. Move down and in **2.** It contracts and flattens **3.** Mucus **4.** Tiny hairs on the mucus making cells **5.** Air out is cleaner, warmer, contains more water vapour and carbon dioxide

The heart and blood vessels

The structure of the heart

The heart is a **muscular pump** that keeps blood flowing through the blood vessels.

The right side of the heart pumps blood **to the lungs** to be oxygenated.

Blood always enters the atria of the heart first. **A** before **V**.

The **heart valves** keep blood flowing **in one direction** and **stop backward flow**.

The left side of the heart pumps blood **around the body**. Blood becomes deoxygenated as it drops off oxygen to the tissues.

The left ventricle has the **thickest muscular wall** because it has to pump blood around the body.

The right ventricle only pumps blood to the lungs.

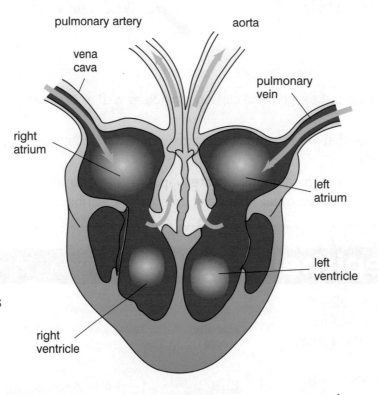

pulmonary artery

aorta

vena cava

pulmonary vein

right atrium

left atrium

left ventricle

right ventricle

The heart gets its own supply from **coronary arteries** which run over its surface.

The double circulation system

This type of blood circulation is called a double circulation because blood flows through the heart **twice** for each complete circulation of the body.

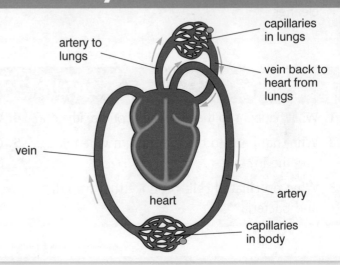

capillaries in lungs

artery to lungs

vein back to heart from lungs

vein

artery

heart

capillaries in body

Blood vessels

Blood is carried around the body in blood vessels.

Veins

Veins carry the blood back to the heart from the body cells and tissues.

They have thinner walls than arteries. They have valves to prevent the blood flowing backwards.

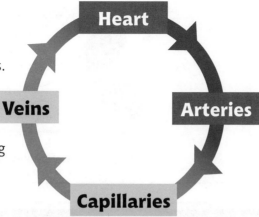

Arteries

Carry blood away from the heart towards the body.

They have thick walls to withstand the high pressure.

Arteries branch into capillaries.

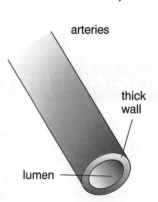

arteries

thick wall

lumen

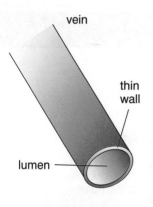

vein

thin wall

lumen

Capillaries

Capillaries are only **one cell thick** to allow **oxygen** and **nutrients** to diffuse out of them and **into** the cells. They take in waste products from cells e.g. **CO_2** and **urea.** They provide a **large surface** for exchange of materials.

longitudinal section of vein

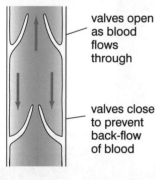

valves open as blood flows through

valves close to prevent back-flow of blood

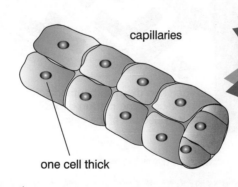

capillaries

one cell thick

Top Tip
The structural differences between arteries, veins and capillaries are worth knowing, especially as they relate to their functions.

Quick Test

1. Why is the heart referred to as a double pump?

2. What is the structural difference between the left side of the heart and the right?

3. Which blood vessels carry blood at high pressure?

4. Which side of the heart contains deoxygenated blood?

5. Name two blood vessels through which blood enters the heart.

6. Name two blood vessels through which blood leaves the heart.

7. What are valves for?

Answers 1. The left side pumps blood to the body and the right side to the lungs **2.** Left side is thicker **3.** Arteries **4.** The right side **5.** Pulmonary vein and vena cava **6.** Pulmonary artery and aorta **7.** To prevent back flow of blood in the heart and the veins

Blood

- Blood is necessary for the transport of oxygen, nutrients and waste products around the body.

- Blood is made up of:

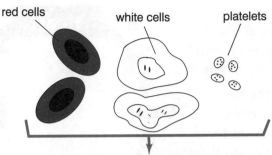

red cells   white cells   platelets

floating in a watery liquid called plasma

Gas exchange

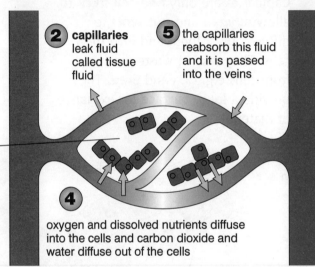

1 **arteries** bring oxygen and dissolved nutrients to the capillaries in the body organs

2 **capillaries** leak fluid called tissue fluid

5 the capillaries reabsorb this fluid and it is passed into the veins

6 the **veins** take the blood back to the heart

3 **tissue fluid** bathes the cells; this is where exchange of substances takes place

4 oxygen and dissolved nutrients diffuse into the cells and carbon dioxide and water diffuse out of the cells

Red blood cells

- Their function is to carry **oxygen** to all the **cells of the body**. They are especially adapted so that they can do this efficiently.
- They have **no nucleus** (more room for oxygen).
- Their shape is a **biconcave disc**, which gives maximum surface area for **absorbing oxygen**.

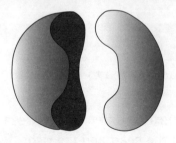

this diagram shows a red blood cell that has been sectioned to show its characteristic shape

Red cells contain **haemoglobin**. Oxygen combines with haemoglobin to form **oxy-haemoglobin**.

In the capillaries, the oxygen is released from the haemoglobin and diffuses into the body cells.

CREDIT

Plasma

- Plasma is a yellow fluid.
- It consists of mainly water, but has many substances dissolved in it. These include **soluble food**, **salts**, **carbon dioxide**, **urea**, **hormones** and **antibodies**.
- Its function is to transport these substances around the body. For example:
  - **soluble food** – from the small intestine to the liver for storage and all cells for respiration
  - **urea** – from the liver to the kidneys for removal
  - **carbon dioxide** – from the cells to the lungs for removal

White blood cells

- Their main function is **defence against disease**.
- They have a **large nucleus**.
- They are **larger** than red blood cells and their shape varies.

white blood cells killing germs by ingesting them

white blood cells killing germs by sending out antibodies

Top Tip
Learn the functions of the four parts of the blood and in particular how the structure of the red blood cell helps it do its job

Platelets

- Platelets are **fragments of cells**.
- Their function is to **clot the blood** so you do not bleed to death if you cut yourself!

Quick Test

1. What are the four main components of blood?
2. What is the function of the white blood cells?
3. What do platelets do?
4. Which types of cell have no nucleus?
5. Name five substances that are dissolved in plasma.

Answers 1. Plasma, red blood cells, white blood cells, platelets **2.** Fight disease **3.** Clot blood **4.** Red blood cells **5.** Urea, hormones, soluble food, carbon dioxide, salts, antibodies

Coordination and the nervous system

Our **nervous system** controls every action we make. The **sense organs** detect information from our surroundings (e.g. sound and light) and send messages to our **brain** and **spinal cord**. The brain and spinal cord sort out this information and send **instructions to the body**, which then responds.

The eye

The eye is the sense organ which **detects light**. Having two eyes allows us to judge distance. Each eye sees a slightly different view.

The brain puts these two views together to form a three-dimensional view which helps in the judgement of distances. This is called **binocular vision**.

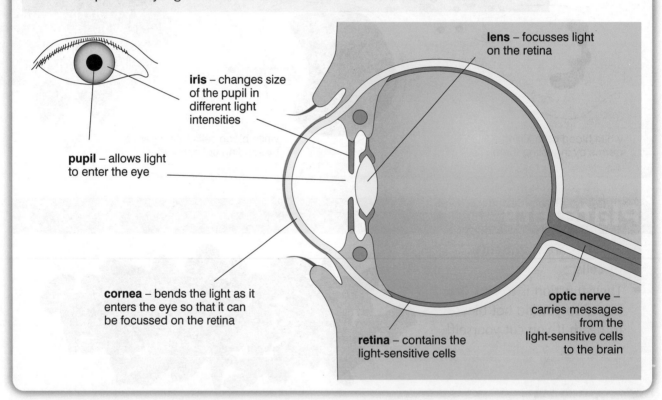

iris – changes size of the pupil in different light intensities

pupil – allows light to enter the eye

cornea – bends the light as it enters the eye so that it can be focussed on the retina

lens – focusses light on the retina

retina – contains the light-sensitive cells

optic nerve – carries messages from the light-sensitive cells to the brain

The iris adjusts to light

Bright light
The iris closes and makes the **pupil smaller**. Less light enters the eye.

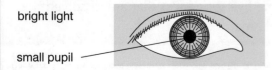

bright light

small pupil

Dim light
The iris opens and makes the **pupil bigger**. More light enters the eye.

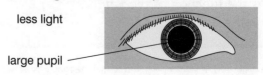

less light

large pupil

The ear

The ear is the **sense organ** used to **detect sound**. Our judgement of where a sound has come from is made more accurate by having two ears.

Top Tip
Make sure you know the main parts of the ear and their functions.

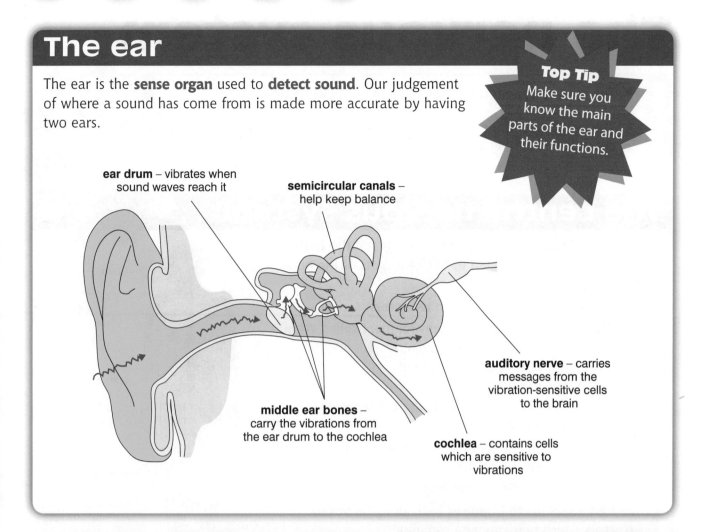

ear drum – vibrates when sound waves reach it

semicircular canals – help keep balance

auditory nerve – carries messages from the vibration-sensitive cells to the brain

middle ear bones – carry the vibrations from the ear drum to the cochlea

cochlea – contains cells which are sensitive to vibrations

Balance

The **semicircular canals** help us to keep our balance. They are three tubes, each at **right angles to each other**. When the head moves, cells in the walls of the canals pick up the movement of fluid in the canals and send messages to the brain.

CREDIT

Quick Test

1. Name the part of the eye that controls the amount of light entering it.

2. What is the name of the hole in the middle of the iris?

3. What happens to the size of the pupil in bright light?

4. What happens to the size of the pupil in dim light?

5. What is the function of the optic nerve?

6. Which part of the eye does light enter through?

7. What enables our judgement of sound to be more accurate?

8. Why is the eardrum important?

9. What is the function of the middle ear bones?

10. Name the nerve that carries messages from the ear to the brain.

11. Why are the semi-circular canals in the ear important?

Answers 1. Iris **2.** Pupil **3.** It gets smaller **4.** It gets larger **5.** Receives nerve impulses from the retina and sends them to the brain **6.** The pupil **7.** Having two ears enables us to detect the direction from which a sound is coming more easily **8.** It vibrates when sound waves reach it **9.** They carry sound vibrations from the ear drum to the cochlea **10.** Auditory **11.** Help you keep your balance

The nervous system

Key Facts

- The nervous system is in charge. It **controls** and **co-ordinates** the parts of your body so that they work together at the right time.
- The nervous system co-ordinates things you do not even think about like breathing and blinking.

Top Tip
Remember: **receptors** are your sense organs; **effectors** are your muscles or glands.

The central nervous system

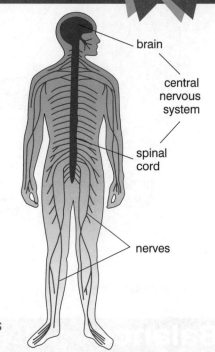

- The **central nervous system** (**CNS**) consists of the **brain** and **spinal cord**, connected to different parts of the body by **nerves**.
- Your body's sense organs contain **receptors**.
- Receptors detect changes in the environment called **stimuli**.
  Nose – sensitive to chemicals in the air
  Mouth – sensitive to chemicals in food
  Ears – sensitive to sound
  Skin – sensitive to touch, pressure and temperature
  Eyes – sensitive to light
- The receptors send messages along nerves to the brain and spinal cord in response to stimuli from the environment.
- The messages are called **nerve impulses**.
- The CNS sends **nerve impulses** back along nerves to **effectors** which bring about a response.
- Effectors are **muscles** that bring about movement, or **glands** that secrete hormones.

Diagram labels: brain, central nervous system, spinal cord, nerves

Nerves

Nerves are made up of nerve cells or **neurones**.

There are three types of neurone.

Neurones have a **nucleus**, **cytoplasm** and **cell membrane**, but they have changed their shape and become specialised.

Nerve impulses travel in **one direction only**.

The **fatty sheath** is for insulation and for speeding up nerve impulses.

A **relay neurone** connects the sensory neurone to the motor neurone in the CNS.

The **sensory neurones** receive messages from the receptors and send them to the CNS.

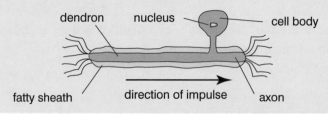

Labels: dendron, nucleus, cell body, fatty sheath, direction of impulse, axon

The **motor neurones** send messages from the CNS to the effectors telling them what to do.

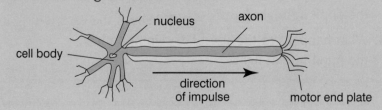

Labels: nucleus, axon, cell body, direction of impulse, motor end plate

Reflex action

A reflex action is a rapid, **automatic response** to a stimulus.

Reflex actions **protect the body** by allowing it to react quickly when in danger.

A reflex action involves a flow of information into and out of the spinal cord. It only involves the brain after the action has taken place.

This is called a **reflex arc** and is shown in the diagram:

1. Stimulus in this example is heat.
2. The receptor is the pain sensor in the skin.
3. The nerve impulse travels along the sensory neurone.
4. The impulse is passed along a **relay neurone** to the motor neurone.
5. The impulse is passed along a motor neurone to the muscle effector in the arm.
6. You move your hand away.

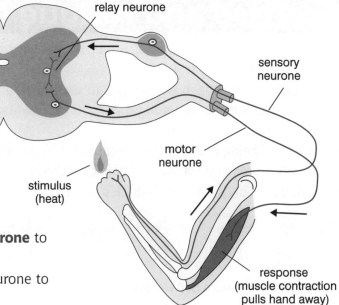

relay neurone

sensory neurone

motor neurone

stimulus (heat)

response (muscle contraction pulls hand away)

> The nervous system always follows this sequence of events:
>
> Stimulus → receptor → sensory neurone → relay neurone → motor neurone → effector → response

The brain

The brain is at the top of the spinal cord and is protected by the skull.

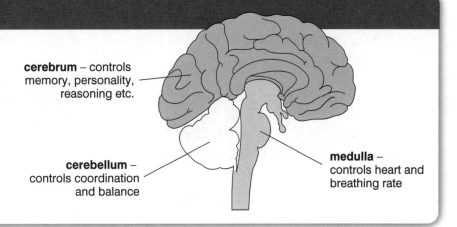

cerebrum – controls memory, personality, reasoning etc.

cerebellum – controls coordination and balance

medulla – controls heart and breathing rate

Quick Test

1. Name the five sense organs.
2. What does the CNS consist of?
3. Which neurone is connected to receptors?
4. Which neurone is connected to the effector?
5. What is a stimulus?
6. Why are reflex actions useful?

Answers 1. Skin, tongue, eyes, ears, nose **2.** The brain and spinal cord **3.** Sensory neurone **4.** Motor neurone **5.** A change in the environment **6.** They protect us from harm.

Changing levels of performance

Exercise and fatigue

- When we run fast, our muscles use up lots of glucose to release energy.
- Sometimes not enough oxygen gets to the muscles, so **respiration without oxygen (anaerobic respiration)** takes place.
- Instead of being broken down to carbon dioxide and water, glucose is broken down into a chemical called **lactic acid**.
- Lactic acid builds up in the muscles and causes **soreness** and **fatigue**.

Top Tip
How lactic acid is produced and how it is removed using the oxygen debt are typical exam questions.

aerobic respiration	glucose + oxygen → carbon dioxide + water + *energy*
anaerobic respiration	glucose → lactic acid + *energy*

At the end of the exercise, we get rid of the lactic acid by breathing rapidly to take in more oxygen. The lactic acid then breaks down to carbon dioxide and water. The volume of oxygen needed to break down the lactic acid is called the **oxygen debt**.

The effects of training

During exercise, the **pulse and breathing rates increase**. This is because your heart is beating faster to carry more glucose and oxygen to the contracting muscles. Your breathing rate also increases. The **recovery time** is the time taken for the pulse, breathing rate and levels of lactic acid to return to normal after the exercise is over.

It **increases heart volume** and the volume of blood sent out during each beat, causing a lower resting pulse rate.

It **increases lung volume** and so increases the amount of oxygen taken in during each breath, causing a lower breathing rate.

Training effects on the body

It produces a **faster recovery time** as lactic acid is removed more quickly.

It **improves circulation** to muscles.

This training effect is shown in the diagram below.

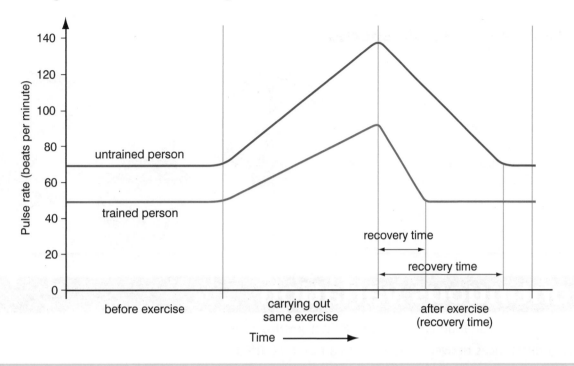

Recovery time

Recovery time can be used as an indication of physical fitness.

A fit person will recover faster than an unfit person.

Top Tip
Make sure you understand how training affects recovery time.

Quick Test

1. What is anaerobic respiration?
2. How is lactic acid produced?
3. What happens when lactic acid builds up in your muscles?
4. How is lactic acid removed?
5. What effect does exercise have on your pulse rate and breathing rate?
6. What is the recovery time?
7. How does training affect recovery time?
8. How does training affect resting heart beat rate?

Answers 1. Breakdown of glucose into lactic acid without oxygen **2.** During vigorous exercise without enough oxygen **3.** It causes soreness and fatigue **4.** By replacing the oxygen debt, fast and deep breathing **5.** They both increase **6.** Time taken for the pulse rate and breathing rate to return to normal **7.** A fit person has a shorter recovery time **8.** A fit person has a slower resting heart beat rate

Variation

Differences within species

- No two members of a species are exactly alike. In other words, variation occurs between individuals.
- There are two types of variation:
  - continuous variation
  - discontinuous variation

Continuous variation

- This type of variation contains **no distinct groups** of individuals.
- Any differences between individuals can be measured.
- Within a large group, any characteristic will vary from **one extreme to another**, e.g. from very short to very tall.

Surveying for continuous variation

A large group can be surveyed for a particular characteristic that is an example of continuous variation. The results can then be displayed as a graph. The shape of the graph may be like the two drawn below.

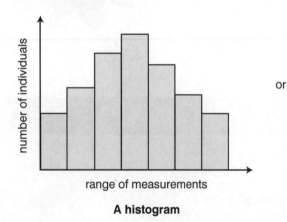

A histogram

or

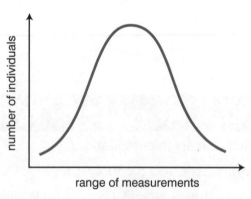

A line graph

Typical examples of characteristics which show continuous variation are **pulse rate** (beats per minute), **height** and **weight**.

Discontinuous variation

In this type of variation, individuals can be divided into two or more **distinct groups**.

Examples of discontinuous variation are blood groups (A, B, AB and O), fingerprint types (loop, arches, whorls, compound), tongue rolling ability.

Surveying for discontinuous variation

A survey of a large group may produce the following types of graphs:

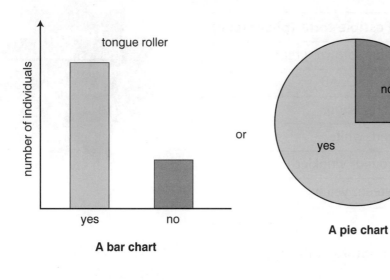

A bar chart

or

A pie chart

Top Tip
When displaying data about continuous or discontinuous variation, make sure you use the correct type of graph or chart.

Quick Test

1. Give two examples of continuous variation.

2. What types of graph are used to display a characteristic that shows continuous variation?

3. Give two examples of discontinuous variation.

4. A population shows four kinds of blood groups. How might the data for this type of variation be displayed?

5. What type of variation contains no distinct groups of individuals?

6. A line graph is a suitable way to represent discontinuous variation. True or false?

7. In discontinuous variation, individuals can be divided into only two distinct groups. True or false?

Answers 1. pulse rate; height; weight; handspan **2.** line graph / histogram **3.** tongue rolling ability; blood groups **4.** bar chart; pie chart **5.** Continuous variation **6.** False **7.** False

What is inheritance?

- All **characteristics** of living things (e.g. hair colour, blood group, colour of flowers) are determined by **genetic information**. This information is carried on the **chromosomes** found in the nucleus of every cell. Each body cell has two matching sets of chromosomes.
- Offspring receive half their genetic information from each parent.

Inherited characteristics

Characteristics can exist in different forms, for example:

Organism	Characteristic	Possible form (phenotype)
human	eye colour	blue, brown, green, grey
	blood group	A, B, O, AB
fruit-fly	wing length	long, short
	eye colour	red, white
pea plant	seed shape	round, wrinkled
	seed colour	green, yellow

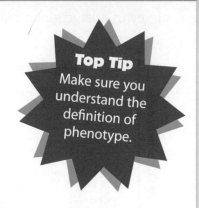

Top Tip
Make sure you understand the definition of phenotype.

The possible forms of each characteristic are called the **phenotypes**.
The phenotype is the **appearance** or **nature** of the organism.

Passing on characteristics

If an organism with a particular phenotype is crossed (mated) with another which shows the same phenotype, it can produce offspring which all show the same phenotype.

If only that phenotype, and no other, appears in all the subsequent generations, the original parent organisms are described as **true breeding**.

e.g. a male black mouse is mated with a female black mouse.

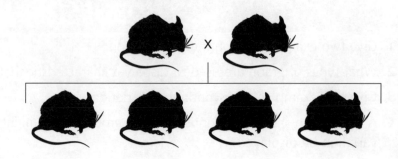

If the offspring all have black coats and all subsequent offspring have black coats, then the parents are **true breeding**.

Dominant and recessive phenotypes

If a true breeding black mouse is crossed with a true breeding 'chocolate' brown mouse the offspring are all black.

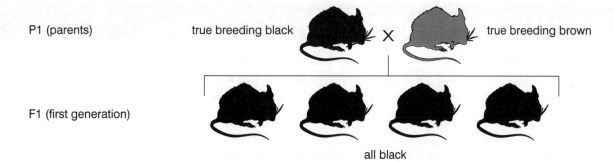

All the offspring are black because this is the 'stronger' phenotype. The phenotype 'black coat' is described as **dominant**. The phenotype 'chocolate coat' is described as **recessive**.

If two of the F1 offspring are crossed with each other then the next generation, the F2 generation, will have some black and some brown haired mice. There will, however, always be more black mice. If large numbers of offspring are produced, the ratio of black (dominant) to chocolate (recessive) will be approximately 3:1.

Quick Test

1. What are inherited characteristics?

2. What is the phenotype of an organism?

3. Name two phenotypes of the characteristic 'eye colour' in the fruit fly.

4. Name two phenotypes of the characteristic 'seed shape' in a pea plant.

5. Name two phenotypes of the characteristic 'eye colour' in people.

More about inheritance

Chromosome sets

The sex cells (**gametes**) contain only one set of chromosomes. The reduction of the number of chromosomes to a single set occurs during gamete formation. When fertilisation takes place, the chromosome set of the egg joins with the chromosome set of the sperm. The nucleus of the fertilised egg (**zygote**) now contains two matching sets of chromosomes. The zygote divides to produce all the other cells of the body. This means that every cell has the same two sets of chromosomes.

When the two sets of chromosomes are examined, they can be arranged in pairs – one of each pair will have come from the mother and the other from the father.

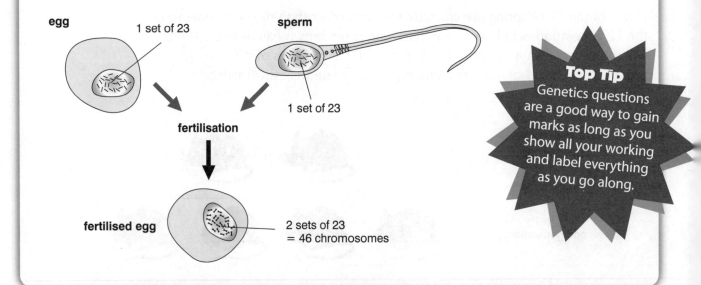

egg
1 set of 23

sperm
1 set of 23

fertilisation

fertilised egg
2 sets of 23
= 46 chromosomes

Top Tip
Genetics questions are a good way to gain marks as long as you show all your working and label everything as you go along.

Genes

Each chromosome carries information in many tiny units called **genes**. Genes determine all the characteristics of an organism.

The gene for each characteristic exists in two or more forms (**alleles**). One form is usually **dominant** and the other is **recessive**. One example in humans is tongue rolling. There are two tongue rolling alleles. One allele allows humans to roll their tongues. The other allele does not. Depending on which combination of these alleles your cells contain, you either can or cannot roll your tongue. The tongue roller allele is dominant to the non-rolling allele.

Phenotype of person	Tongue roller	Tongue roller	Non-tongue roller
Alleles contained in body cells	tongue rolling allele tongue rolling allele	tongue rolling allele non-rolling allele	non-rolling allele non-rolling allele

A true breeding cross

The diagram below shows how the coat colour alleles are passed on when true breeding mice are crossed.

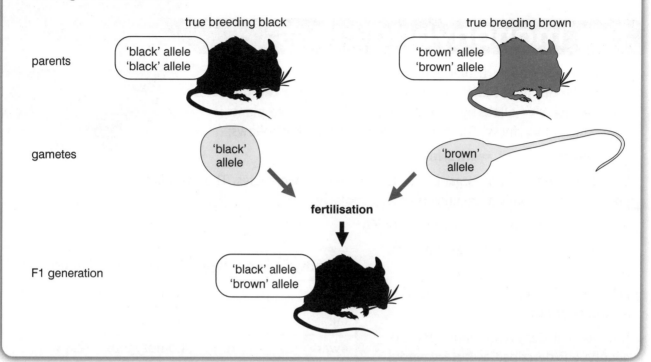

parents — true breeding black — 'black' allele / 'black' allele

true breeding brown — 'brown' allele / 'brown' allele

gametes — 'black' allele — 'brown' allele

fertilisation

F1 generation — 'black' allele / 'brown' allele

Definitions

- **Recessive** means it is the weaker allele.
- **Dominant** means it is the stronger allele.
- The **genotype** is the type of alleles an organism carries; the genotype of the black mice could be BB or Bb. Although the genotypes are different, the mice are still black because black is dominant.
- The **phenotype** is what the organism physically looks like, the result of what genotype it has.

Quick Test

1. What is the genotype of an organism?
2. What is a dominant gene?
3. What is a recessive gene?
4. What is a phenotype?
5. Gametes contain two sets of chromosomes. True or false?
6. A zygote contains two matching sets of chromosomes. True or false?

Answers 1. The type of genes (alleles) an organism carries for a characteristic **2.** The stronger allele (gene) that expresses itself in the phenotype **3.** The weaker allele (gene) that can only express itself in the absence of a dominant gene **4.** The phenotype is what the organism physically looks like, the result of what genotype the organism has **5.** False **6.** True

How genes are inherited

The monohybrid cross

Top Tip
It is crucial to understand the terms homozygous and heterozygous if you are to be successful in working out genetic crosses.

The two forms of a particular gene (allele) can be represented by letters. The **dominant allele** is always represented by a **capital letter** and the **recessive allele** by a **small letter**. Tallness in pea plants is dominant to dwarfness. These alleles are represented as follows:

T = tallness   **t = dwarfness**

Each characteristic of an organism is determined by two alleles. The two alleles present in an organism are known as its **genotype**.

A tall pea plant has the genotype **TT** or **Tt**.

A dwarf pea plant has the genotype **tt**.

If the two alleles are the same, (**TT** or **tt**), the genotype is described as **homozygous**.

If the two alleles are different (**Tt**), the genotype is described as **heterozygous**.

Phenotype	Genotype	Description of genotype
tall	TT	homozygous dominant
tall	Tt	heterozygous
dwarf	tt	homozygous recessive

The simplest genetic cross involves one characteristic and is called a **monohybrid cross**.

Parents' phenotypes	green stem tomato	X	white stem tomato
Parents' genotypes	GG		gg
Gametes	G	fertilisation	g
F1 genotype		Gg	
F1 phenotype		all green	

The next step might involve crossing two individuals from the F1 generation.

F1 phenotype	green stem tomato	X	green stem tomato
F1 genotype	Gg		Gg
Gametes	G or g		G or g

Punnett square

A **Punnett square** is then used to show the possible ways in which these two sets of gametes could combine during fertilisation.

The 3:1 ratio is only likely to occur when large numbers of offspring are produced. This is because the fusion of gametes at fertilisation is random and is a matter of chance (like tossing a coin).

F2 genotypes

female gametes

male gametes		G	g
	G	**GG** green	**Gg** green
	g	**gG** green	**gg** white

F2 phenotype ratio   3 green : 1 white

The inheritance of sex

- There are 23 pairs of human chromosomes.
- The **23rd pair is the sex chromosomes**; they determine whether you are a boy or a girl. All the other chromosomes contain information for your characteristics.
- If you are a male, one of the sex chromosomes is **shorter** than the other: this is the **Y chromosome**.
- Females have two **X chromosomes** that are the **same size**.
- The female ovary will produce only X chromosomes.
- The male testis will produce half X chromosome sperms and half Y chromosome sperms.

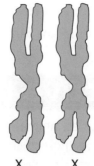
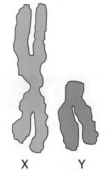

X	Y	X	X

male sex chromosomes   female sex chromosomes

- During fertilisation the egg may join with either the X sperm or the Y sperm.

We can show this in a Punnett square diagram:

- The diagram shows that each time a couple have a child, there is a **50% chance** that it will be **male** and a **50% chance** that it will be **female**.
- The 50% chance of a boy or a girl is only a probability, all of the children could be girls or all boys.

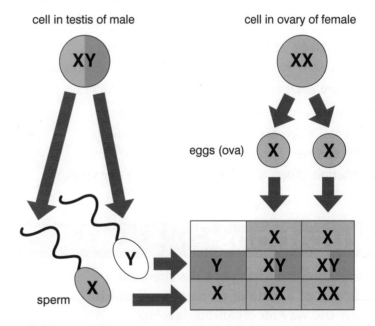

cell in testis of male   cell in ovary of female

eggs (ova)

sperm

	X	X
Y	XY	XY
X	XX	XX

Quick Test

1. What is a homozygous genotype?

2. What is a heterozygous genotype?

3. What is a monohybrid cross?

4. How many types of gamete do you get from a heterozygous tall plant?

5. If two heterozygous tall plants are crossed, what is the ratio of offspring in the F_1 generation?

6. What is the genotype of a human male?

7. What is the genotype of a human female?

8. Each time a couple have a child, what are the chances that it will be a female?

Genetics and society

What is a species?

A species is a group of living things that are so similar to each other that:

a. they are **able to interbreed** and produce offspring;

b. the **offspring are fertile** (able to produce offspring of their own).

Selective breeding

- For centuries, animal and plant breeders have tried to improve their stock and their crops (e.g. dairy farmers want to have cows which produce large quantities of milk).

- Improved characteristics can be obtained by selective breeding.

- Breeders select and breed those varieties of animals and plants with characteristics that are useful.

Examples of selective breeding

- Aberdeen Angus cattle are bred for beef production.
- Jersey cows are bred for high milk production.
- Poultry are bred to grow more quickly.
- Cereal crops are bred to be more resistant to disease.
- Merino sheep are bred to produce better wool.

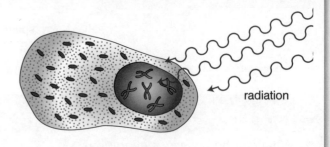

Top Tip
You are often asked to give examples of selective breeding in exam questions.

Mutations

A mutation is a change in the chemical structure of a gene or chromosome, which alters the way an organism develops.

The change may happen for no reason or there might be a definite cause.

Mutations occur naturally in the environment; for example, new strains of the flu virus are always appearing.

radiation

The chances of mutation can be increased by:

- **Ionising radiation**, such as alpha, beta and gamma radiation. This radiation may damage the DNA inside cells.

- **X-rays** and **ultraviolet light**, which may alter the genes in such a way that when they divide to replace themselves, uncontrollable growth is produced which develops into tumours and cancer.

- **Certain chemicals** called **mutagens**. Cancer-causing chemicals are called carcinogens and are found, for example, in cigarette smoke.

Amniocentesis

A technique called **amniocentesis** can be used to **detect chromosome changes** in an embryo.

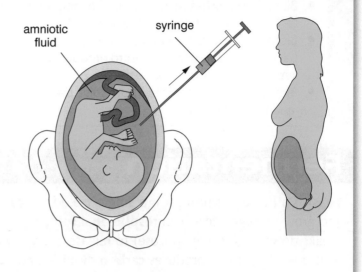

- A sample of amniotic fluid is removed using a syringe.
- The fluid contains cells from the baby's skin and chemicals from the lungs and urine.
- The cells can be grown in the laboratory and then examined to see if there are any defects in the chromosomes.
- The fluid can be tested to see if it contains certain chemicals which may indicate a chromosome abnormality.

Down's syndrome

- A **chromosome mutation** is a change in the number of chromosomes in the cell.
- When cells divide to form gametes (sex cells), they share out equally the number of chromosomes, so each has 23.
- Occasionally an extra chromosome goes into one cell and the fertilised egg will have 47 chromosomes instead of 46.
- The child has **three chromosome number 21s** instead of two in their cells and will have **Down's syndrome**.
- The child can live a relatively normal life, but will be mentally disabled and be more susceptible to some diseases.

Quick Test

1. What is a species?
2. What is selective breeding?
3. Give an example of selective breeding in plants.
4. Give an example of selective breeding in animals.
5. What is a mutation?
6. State two things that can increase mutation rate.
7. Name the technique used to detect chromosome change in the embryo.
8. What causes Down's Syndrome?

Answers 1. A group of living things that are able to interbreed and produce fertile offspring **1.** Breeders select and breed those varieties of animals and plants with characteristics that are useful **2.** Cereal crops are bred to be disease resistant **3.** Aberdeen Angus cattle bred for beef production **4.** A change in the genetic material in chromosomes **5.** X-rays, radiation **6.** Amniocentesis **7.** One extra chromosome 21

Biotechnology

Living factories

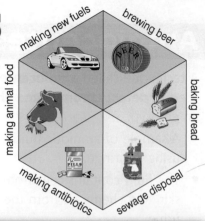

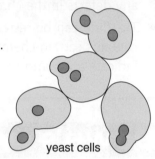

making new fuels · brewing beer · baking bread · sewage disposal · making antibiotics · making animal food

Key Facts

- Biotechnology has many uses in the modern world; its techniques have been used for a long time.
- Biotechnology uses living cells from plants, animals, bacteria and fungi (microbes) to produce useful substances or dispose of waste.
- Biotechnology may in the future solve the world's health, and food problems.

Fermentation

Yeast is a living organism. It is a **single-celled fungus** which can feed and grow on sugar. Yeast can **respire anaerobically**. This means that it breaks down sugars (e.g. glucose) to release energy without using oxygen. As it does this, it also produces **carbon dioxide** and **ethanol (alcohol)**.

This process is called **fermentation** and is summarised in the word equation below.

$$\text{glucose} \longrightarrow \text{carbon dioxide} + \text{alcohol} + \text{energy}$$

yeast cells

Bread

Yeast is important in baking because the **carbon dioxide produced makes the bread rise**. Baking the bread finally kills the yeast and cooks the dough.

Beer

- Fermenting barley makes beer: it is flavoured with a substance called hops.
- Barley seeds germinate to make sugar.
- Yeast is added to the **fermenter**. It uses the sugar from barley to make **alcohol** and **carbon dioxide**. The alcohol is filtered and pasteurised to kill unwanted organisms and then bottled.

Batch processing

Allowing yeast to ferment sugar can be done on a large scale. Batch processing is a technique used by commercial brewers. A large reactor vessel (**a fermenter**) is filled with the necessary raw materials and given the best conditions to promote fermentation.

After the fermenter has been set up, the system is closed and left until fermentation is complete. Then the products can be collected and purified.

Yeast grows best in warm conditions where the temperature is between 10° and 18° C, with an adequate supply of glucose and oxygen and a pH of 7. An absence of other micro-organisms (sterile conditions) is essential as these may slow down the process.

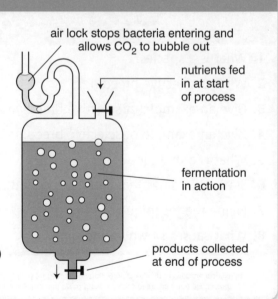

air lock stops bacteria entering and allows CO_2 to bubble out

nutrients fed in at start of process

fermentation in action

products collected at end of process

The malting of barley

To make beer, brewers use barley as food for yeast. However, barley grains contain starch and not the simple sugar that yeast can feed on. The barley must be allowed to **germinate in moist, warm conditions** in a process called **malting**. During this time, enzymes present in the barley break down the starch into the sugar **maltose** on which the yeast can feed.

$$\text{starch} \xrightarrow{\text{enzymes in barley grains}} \text{maltose} \xrightarrow{\text{yeast}} \text{carbon dioxide} + \text{alcohol} + \text{energy}$$

Other fermentation processes

Fresh milk contains a sugar called **lactose**. When bacteria feed on this sugar, they break it down in a fermentation process and produce **lactic acid**. This acid makes the milk taste **sour**. The pH of the milk becomes more and more acidic as more lactose is fermented.

$$\text{lactose} \xrightarrow{\text{bacteria}} \text{lactic acid}$$

Yoghurt

Yoghurt is produced by the fermentation of low-fat milk. Making yoghurt uses two species of bacteria, **Streptococcus** and **Lactobacillus**. The bacteria use the milk as a food source and decrease pH by producing lactic acid from the lactose sugar in milk. A pH of 4.4 causes the proteins to coagulate (thicken), so making thick yoghurt.

Cheese

Rennet is added to milk. The milk curdles to form **curds** (solid) and **whey** (liquid). The curds are **separated** from the whey. **Salt** is added to the curds for **flavour** and to **prevent the growth of unwanted bacteria**. These curds are **pressed in a mould** to form hard cheese.

Top Tip
Learn examples of foods that are produced as a result of biotechnology.

Quick Test

1. What is biotechnology?

2. What kind of organism is yeast?

3. Write down the word equation for fermentation using yeast.

4. Why is yeast important in the baking industry?

5. Name the useful product of fermentation that is important in the brewing industry.

6. Name the large vessel used by commercial brewers when making beer.

7. Write the word equation for the malting process in barley grains.

Answers 1. The use of living cells to produce useful substances or dispose of waste **2.** A single-celled fungus **3.** Glucose ⟶ carbon dioxide + alcohol + energy **4.** The carbon dioxide produced causes the dough to rise **5.** Alcohol **6.** A fermenter **7.** Starch ⟶ maltose

Problems with waste

> **Key Fact**
>
> **The problem of waste**
> In nature and in industry, **waste** products are continually being produced. If they were not recycled in some way, the world would accumulate more and more unusable waste.

Working with microbes

Microbes are tiny living things that can only be seen using a microscope. They include bacteria, viruses and fungi.

Microbes can be useful, but they can also be harmful.

> **Top Tip**
> Make a list of the ways in which useful and harmful microbes can affect our lives.

Precautions with microbes

Certain precautions must be taken when working with microbes to ensure unwanted microbes are not able to grow and cause disease.

1. **Wash** hands and lab bench.

2. Always use **sterile** equipment.
3. Never open dishes containing microbes.
4. Always dispose of dishes containing microbes by using high temperatures.

Contamination

Contamination is the presence of **unwanted**, possibly **harmful microbes**. Many manufacturing processes (e.g. brewing) have to take special **precautions** to ensure that all equipment is **sterile**. After every batch of beer is brewed the equipment is sterilised by using steam heat and chemicals. This is to destroy resistant fungi and bacteria which other methods of sterilisation do not kill.

Microbes and decay

In nature, **decay** ensures that minerals locked up in the dead remains of plants and animals are **recycled**.

After a plant or animal dies, its tissues decay. It is **microbes** that cause the decay. Microbes use dead material as their food source to obtain energy and building materials to stay alive and grow. These microbes, called **decomposers**, can only use organic (natural) substances. (Substances such as plastic do not decay easily.) The **nitrogen** and **carbon cycles** (see pages 10–13) illustrate nature's way of turning waste into useful products; both involve microbes.

CREDIT

Problems with untreated sewage

Untreated sewage contains faeces, detergents, food fragments and bacteria, which can cause great damage if dumped into a river.

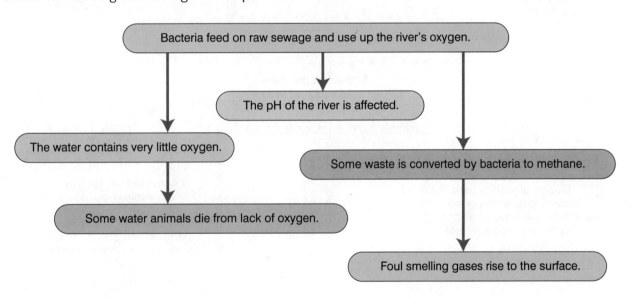

Bacteria feed on raw sewage and use up the river's oxygen.

The pH of the river is affected.

The water contains very little oxygen.

Some waste is converted by bacteria to methane.

Some water animals die from lack of oxygen.

Foul smelling gases rise to the surface.

Some of the microbes in sewage cause diseases such as **dysentery**, **typhoid**, **cholera** and **food poisoning**.

Quick Test

1. What are microbes?

2. Why must precautions be taken when working with microbes?

3. What term is used to describe the presence of unwanted microbes?

4. What are decomposers?

5. List three things found in untreated sewage.

6. List three things that can happen if sewage is dumped in a river.

Answers: 1. Tiny living organisms eg bacteria, viruses, fungi that can only be seen using a microscope. **2.** To ensure that unwanted microbes are not able to grow and cause disease. **3.** Contamination. **4.** Microbes which can cause decay by feeding on dead organic material. **5.** Faeces, detergents, food fragments, bacteria. **6.** pH of river changes, dissolved oxygen level falls, aquatic animals die, methane gas may be produced by bacteria

Profit from waste

Sewage treatment

The treatment of sewage involves the conversion of harmful materials to harmless products by the use of decay microbes.

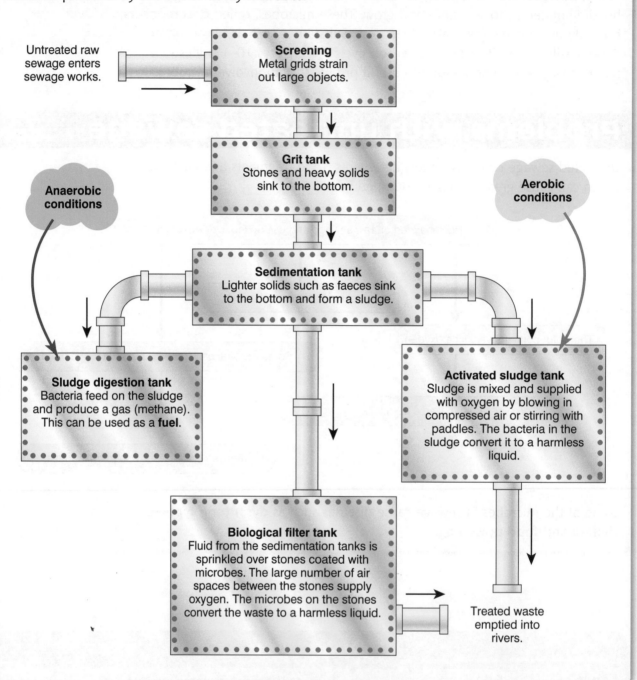

Untreated raw sewage enters sewage works.

Screening
Metal grids strain out large objects.

Grit tank
Stones and heavy solids sink to the bottom.

Anaerobic conditions

Aerobic conditions

Sedimentation tank
Lighter solids such as faeces sink to the bottom and form a sludge.

Sludge digestion tank
Bacteria feed on the sludge and produce a gas (methane). This can be used as a **fuel**.

Activated sludge tank
Sludge is mixed and supplied with oxygen by blowing in compressed air or stirring with paddles. The bacteria in the sludge convert it to a harmless liquid.

Biological filter tank
Fluid from the sedimentation tanks is sprinkled over stones coated with microbes. The large number of air spaces between the stones supply oxygen. The microbes on the stones convert the waste to a harmless liquid.

Treated waste emptied into rivers.

Aerobic conditions

Many different types of microbe are used in sewage treatment to ensure the complete breakdown of all the different substances present in the sewage. Complete breakdown is only possible in **aerobic conditions**, i.e. when oxygen is present, for aerobic respiration. Without oxygen, sewage is only partially broken down, leaving some harmful products.

Upgrading waste

Many manufacturing processes produce **organic waste products**. These can be fed to microbes which convert them to products that are useful to people and other animals.

> The advantage is that unwanted waste products can be converted to products with a **high energy** or **protein** value. Consequently, they have a **higher economic value** than the original waste.

 CREDIT

Top Tip

Make sure you learn all the stages involved in sewage treatment and make a list of useful substances that result from upgrading waste.

Fuels from microbes

When microbes grow on fresh manure, they produce **methane gas**. When yeast feeds on sugar, it produces **alcohol**. Both the methane and the alcohol can be used as **fuels**.

These two fuels are described as **renewable** energy sources, whereas fossil fuels (coal, oil and gas) are **non-renewable** energy sources. There are advantages to using renewable energy sources rather than fossil fuels. They are less harmful to the environment, easy and cheap to obtain, and will never be used up.

Protein-rich foods

Microbes can reproduce very rapidly by dividing into two. Given food, water and heat, one bacterium can result in many hundreds of bacteria within a few hours. Industry makes use of fast-growing bacteria to produce protein-rich foods, as a high percentage of a bacterium is protein.

Bacteria can be grown, harvested and dried to form a protein-rich powder called **single-celled protein** which is used as animal feed.

Mycoprotein (Quorn®) is made from a fungus and is a useful source of protein for vegetarians.

Quick Test

1. During sewage treatment what happens in the sludge digestion tank?

2. Name two ways in which oxygen is added to the activated sludge tank.

3. How is oxygen supplied in the biological filter tank?

4. Why are many types of bacteria used in sewage treatment?

5. What would happen if oxygen was not present during sewage treatment?

6. Give two examples of useful products that result from upgrading waste.

Answers 1. Bacteria feed on sludge and produce methane gas. **2.** Blowing compressed air, stirring with paddles. **3.** The large number of air spaces between the stones supply oxygen. **4.** To ensure the complete breakdown of all the different substances found in sewage. **5.** The sewage would only be partly broken down, leaving some harmful products. **6.** Methane gas from manure used as a fuel; alcohol from sugar used as a fuel; protein-rich animal feed.

Biotechnology

Reprogramming microbes

Genetic engineering

The normal control of activity in bacteria depends on chromosomes. Genetic engineering involves **transferring a gene from one organism into a bacterium**. The bacterium then makes the chemical for which the transferred gene contains the information. The **bacterium reproduces rapidly**. All the new bacteria will also make the chemical, meaning large quantities of a chemical, such as insulin, can be made.

Top Tip
Make sure you know the main points about genetic engineering and immobilisation as these are important techniques used in biotechnology.

Products of genetic engineering

Insulin

Insulin is a hormone that controls blood sugar level. People with diabetes cannot produce insulin. Until recently, sufferers used sheep or pig insulin, but some people are **allergic** to it. Genetic engineering reprogrammes bacteria to produce insulin identical to human insulin.

Producing insulin

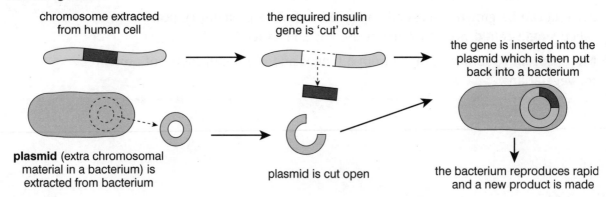

chromosome extracted from human cell

the required insulin gene is 'cut' out

the gene is inserted into the plasmid which is then put back into a bacterium

plasmid (extra chromosomal material in a bacterium) is extracted from bacterium

plasmid is cut open

the bacterium reproduces rapid and a new product is made

Antibiotics

Antibiotics, which can kill some harmful bacteria, are products of genetic engineering. Penicillin is produced by a fungus. Different antibiotics are produced, as a single antibiotic is not always effective against bacteria. This is because bacteria can become resistant to a particular antibiotic.

Advantages of genetic engineering

Genetic engineering and **selective breeding** both result in **changes to a genotype**. Genetic engineering is **faster** and produces the organism with the required genotype **immediately**. Selective breeding cannot manipulate the chromosomes themselves. It is a **long process** which **does not always produce the organism required**.

Biological detergents

Biological detergents contain **enzymes** produced by bacteria. They break down stains, such as grass, blood and egg. They are more effective at removing stains at **low temperatures** (40°C) than non-biological detergents. This prevents **damage to fabrics** and reduces fuel costs.

Immobilisation

Immobilisation **fixes enzymes** onto substances such as jelly or glass beads so that they can be used **repeatedly** without having to separate them continually from the product of their action. Whole cells, such as yeast, can be immobilised this way.

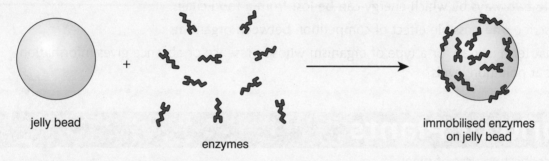

jelly bead

enzymes

immobilised enzymes on jelly bead

CREDIT

Continuous flow processing

Continuous flow processing is where immobilised enzymes are placed in a fermenter. Nutrients are continuously fed in and the end products continuously collected and purified. This method is **more productive** and **less expensive** than batch processing (used in brewing).

Top Tip
Exam questions often ask you to compare batch processing with continuous flow processing.

Continuous flow processing

nutrients fed in continuously

immobilised enzymes in reactor

products collected continuously

CREDIT

Quick Test

1. What is genetic engineering?

2. Name the product of genetic engineering that is used to treat diabetes.

3. Name an antibiotic produced by a fungus.

4. What is a biological detergent?

5. What is an immobilised enzyme?

Practice questions

Well done! You've made it through each of the Standard Grade Biology topics.
Now, try answering the following questions to see how much you can recall.
Check your answers on pages 94 and 95.

Biosphere

1. Write a definition for each of the following.
   a. an ecosystem   **b.** a population

2. What term is used to describe a diagram that shows the numbers of organisms present at successive levels in a food chain?

3. State **two** ways by which energy can be lost from a food chain.

4. Describe **one** possible effect of competition between organisms.

5. What term is used for a type of organism whose presence or absence gives information about pollution levels?

World of Plants

1. Describe **two** uses of plants.

2. Name **two** methods of pollination.

3. Describe the features of a flower which make it suitable for insect pollination.

4. Describe what happens to the following after pollination.
   a. ovary   **b.** ovule

5. Describe **two** methods of seed dispersal.

6. Write the word equation for photosynthesis.

Animal Survival

1. Describe **two** features of the small intestine which help in the absorption of the products of digestion.

2. Describe **one** function of the large intestine.

3. Where are the following sex cells produced?
   a. sperm   **b.** eggs

4. Describe what happens to the zygote after fertilisation.

5. Name a substance which is filtered from and then reabsorbed into the blood in the kidney.

6. Urine can account for 50% of all the water lost from the body.
   Name **one** other way in which water can be lost.

7. Name the waste product which would build up in the body if the kidneys were damaged.

Investigating Cells

1. Name **two** features which are found only in plant cells.
2. Describe how cells can be treated so that the cell contents become more visible.
3. Cells increase in numbers by cell division. Name the part of the cell that controls cell division.
4. Explain why enzymes are required in living cells.
5. Name **two** enzymes found in the digestive system and state what they break down.
6. Name **two** types of enzyme reactions.
7. Write the word equation for aerobic respiration.
8. Which of the main types of food components (carbohydrates, fats or proteins) contain the most chemical energy per gram?

Body in Action

1. Name **two** types of movable joint and describe their range of movements.
2. Name the four chambers in the human heart.
3. Describe the function of valves in the heart.
4. Name the type of blood vessel where exchange of materials takes place between the blood and body cells.
5. Which parts of the brain are concerned with
   a. balance   **b.** control of breathing and heart rates?

Inheritance

1. State the type of variation shown by the following characteristics:
   a. blood groups   **b.** height
2. The gene (B) for black coat colour is dominant to the gene (b) for brown coat colour. What term is used to describe the genotype Bb?
3. What is the general term used for sex cells?
4. A human baby inherits an X chromosome from its father.
   a. State the genotype of the baby.   **b.** State the sex of the baby.

Biotechnology

1. Name **two** manufacturing industries that depend on the activities of yeast.
2. Give **one** advantage of using biological detergents in the home, rather than non-biological detergents.
3. Explain why a range of antibiotics is needed in the treatment of bacterial diseases.
4. What is meant by the word 'immobilised' in connection with enzymes?
5. During the production of insulin, a piece of a human chromosome is transferred into a bacterial cell. What name is given to this procedure?

Answers

Biosphere

1. **a.** part of the biosphere / made up of community and habitats.

   b. all the organisms of one type or species in an ecosystem

2. Pyramid of numbers

3. Heat, movement, in faeces, urine

4. Some organisms may die due to lack of food / some may not find a suitable breeding site / some may fail to breed

5. Indicator species

World of Plants

1. Food source / raw material / medicine

2. Wind pollination; insect pollination

3. Large / brightly coloured / scent / nectar

4. Ovary develops into fruit / ovules develop into seeds

5. By wind / animal internal / animal external

6. Carbon dioxide + water $\xrightarrow[\text{sunlight}]{\text{chlorophyll}}$ glucose + oxygen

Animal Survival

1. Large surface area of villi / thin walls / dense blood capillary system

2. To reabsorb water from waste / to move waste towards anus

3. **a.** testes of male   **b.** ovaries of female

4. Divides repeatedly to form a ball of cells which implants in the uterus wall

5. Glucose

6. In faeces; sweating; breathing out

7. Urea

Investigating Cells

1. Cell walls; large vacuole; chloroplasts

2. Add stains / dyes e.g. iodine solution

3. Nucleus

4. They are catalysts i.e they speed up the rate of chemical reactions

5. Amylase breaks down starch / pepsin breaks down proteins / lipase breaks down fats
6. Breakdown and synthesis
7. Glucose + oxygen → carbon dioxide + water + energy
8. Fats

Body in Action

1. Hinge – movement in **one** plane only
   Ball and socket – movement in all planes
2. Right atrium, left atrium, right ventricle, left ventricle
3. Prevent backward flow of blood
4. Capillary
5. **a.** cerebellum   **b.** medulla

Inheritance

1. **a.** discontinuous **b.** continuous
2. Heterozygous
3. Gametes
4. **a.** XX   **b.** Female

Biotechnology

1. Brewing, baking
2. Low temperature washes / saving on fuel / less likely to damage fabrics
3. Some bacteria may be resistant to a certain type of antibiotic.
4. They cannot move as they are attached to a jelly or glassbead.
5. Genetic engineering

Exam practice

Congratulations on completing your revision using this Success Guide for Standard Grade Biology.

If you answered most of the questions above correctly, you will be ready to practise for your exam using Leckie and Leckie's Past Papers for Standard Grade Biology. These will give you an idea of how the exam will be, and will allow you to practise in 'real' exam time. Good luck!

Standard Grade Biology

Index